Applications of Novel Scintillators for Research and Industry

(ANSRI 2015)

Journal of Physics: Conference Series Volume 620

Dublin, Ireland
12 – 14 January 2015

Editors:

Oliver Roberts
Lorraine Hanlon
Sehila McBreen

ISBN: 978-1-5108-0755-6
ISSN: 1742-6588

Printed from e-media with permission by:

Curran Associates, Inc.
57 Morehouse Lane
Red Hook, NY 12571

Some format issues inherent in the e-media version may also appear in this print version.

Printed by Curran Associates, Inc. (2015)

For permission requests, please contact the Institute of Physics
at the address below.

Institute of Physics
Dirac House, Temple Back
Bristol BS1 6BE UK

Phone: 44 1 17 929 7481
Fax: 44 1 17 920 0979

techtracking@iop.org

Additional copies of this publication are available from:

Curran Associates, Inc.
57 Morehouse Lane
Red Hook, NY 12571 USA
Phone: 845-758-0400
Fax: 845-758-2634
Email: curran@proceedings.com
Web: www.proceedings.com

TABLE OF CONTENTS

Journal of Physics: Conference Series **620** (2015) 011001 doi:10.1088/1742-6596/620/1/011001

Preface

Scintillator detectors are used widely in the field of γ- and X-ray spectroscopy, particularly in the mid 1900s when the invention of NaI(Tl) by nobel laureate Robert Hofstadter in 1948, spurred the creation of new scintillator materials. In the development of such new scintillators, important characteristics such as its intrinsic efficiency, position sensitivity, robustness, energy and timing response, light output, etc, need to be addressed. To date, these requirements cannot be met by a single type of scintillator alone and therefore the development of an "ideal" scintillator remains the holy grail of nuclear instrumentation. Consequently, the last two decades have seen significant progress in the development of scintillator crystals, driven largely by technological advances. Conventional inorganic scintillators such as NaI(Tl) and BGO are now being replaced with better, novel organic, inorganic, ceramic and plastic scintillators offering a wider variety of options for many applications.

The workshop on the Applications of Novel Scintillators in Research and Industry was held at University College Dublin in January 2015 and covered a wide range of topics that characterise modern advances in the field of scintillator technology. This set of proceedings covers areas including the growth, production and characterisation of such contemporary scintillators, along with their applications in various fields, such as; Medical Imaging; Defence/Security; Astrophysics; and Nuclear/Particle Physics. We would like to thank all those who presented their recent results on their research at the workshop. These proceedings atest to the excitement and interest in such a broad field, that pervades the pursuit of the development of novel materials for future applications.

We would also like to thank Professor Luigi Piro, for giving an interesting public talk during the conference, and to the Institute of Physics Ireland Group for supporting the event. We thank ORTEC for sponsoring the Poster Prize and Science Foundation Ireland, Scionix, and the Institute of Physics Nuclear Physics, Nuclear Industry and Materials and Characterisation groups for sponsoring the workshop. We also thank University College Dublin for use of the venue and general support that was given during the workshop.

Finally, I would like to thank the members of the Local Organising Commitee for their help in making the workshop a success, and to Dr. Sheila McBreen and Professor Lorraine Hanlon for their help and support, particularly in putting these proceedings together. In addition to this we would also like to recognise the contributions made by the Scientific Orgainising Committee for their advice regarding the organisation of the programme, which was engaging and stimulated a lot of interest amongst the audeince.

Dr. Oliver J. Roberts

Workshop Chair
On behalf of the organising commitees.

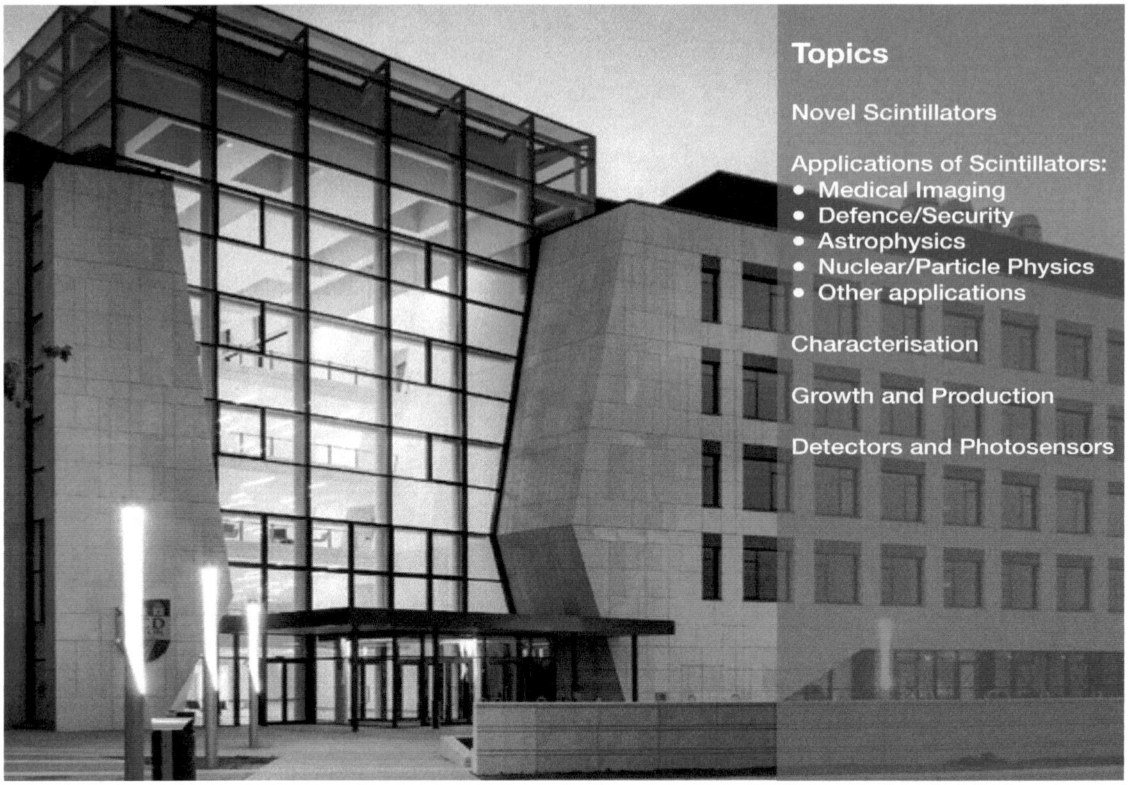

Applications of Novel Scintillators for Research and Industry

12-14 January 2015 — University College Dublin, Ireland

http://ssmr.ucd.ie/ansri2015

Local Organising Committee

Oliver Roberts (Chair)
Sinéad Hales
Lorraine Hanlon
Antonio Martin-Carrillo
Sheila McBreen
David Murphy
Alexey Uliyanov

Invited Speakers

Nerine Cherepy, LLNL, US
David Jenkins, York, UK
Paul Lecoq, CERN, CH
Marek Moszyński, NCBJ, PL
Francesco Quarati, Delft, NL
Stefano Sanvito, TCD, IRE
Paul Schotanus, SCIONIX, NL
Brent Wagner, Georgia Tech, US

Science Organising Committee

Sheila McBreen (Chair), UCD
Franco Camera, INFN-Milano
Nerine Cherepy, LLNL
Lorraine Hanlon, UCD
Paul Lecoq, CERN
Julie McEnery, NASA
Oliver Roberts, UCD
Anant Setlur, GE
Brian Shortt, ESA
Kenneth Stanton, UCD

IOP | Institute of Physics
Nuclear Physics Group
Nuclear Industry Group
Materials and Characterisation Group

ORTEC® AMETEK® ADVANCED MEASUREMENT TECHNOLOGY

sfi Fondúireacht Eolaíochta Éireann Science Foundation Ireland

UCD DUBLIN

SCIONIX

Applications of Novel Scintillators for Research and Industry (ANSRI 2015) Group Photo

Journal of Physics: Conference Series **620** (2015) 011003 doi:10.1088/1742-6596/620/1/011003

Edited by:

Oliver Roberts
Lorraine Hanlon
Sheila McBreen

Local Organising Committee:

Oliver Roberts (**Chair**)
Antonio Martin-Carrillo
Lorraine Hanlon
Sheila McBreen
Alexey Uliyanov
David Murphy
Sinéad Hales

Scientific Organising Committee:

Sheila McBreen, (UCD, Ireland) **(Chair)**
Franco Camera (INFN-Milano, Italy)
Nerine Cherepy (LLNL, USA)
Jarek Glodo (RMD, USA)
Lorraine Hanlon (UCD, Ireland)
Paul Lecoq (CERN, Switzerland)
Julie McEnery (NASA, USA)
Oliver Roberts (UCD, Ireland)
Anant Setlur, (GE, USA)
Brian Shortt, (ESA, the Netherlands)
Kenneth Stanton (UCD, Ireland)

Journal of Physics: Conference Series **620** (2015) 011004 doi:10.1088/1742-6596/620/1/011004

Peer review statement

All papers published in this volume of *Journal of Physics: Conference Series* have been peer reviewed through processes administered by the proceedings Editors. Reviews were conducted by expert referees to the professional and scientific standards expected of a proceedings journal published by IOP Publishing.

Novel scintillators and silicon photomultipliers for nuclear physics and applications

David Jenkins
Department of Physics, University of York, York YO10 5DD, UK
E-mail: david.jenkins@york.ac.uk

Abstract. Until comparatively recently, scintillator detectors were seen as an old-fashioned tool of nuclear physics with more attention being given to areas such as gamma-ray tracking using high-purity germanium detectors. Next-generation scintillator detectors, such as lanthanum bromide, which were developed for the demands of space science and gamma-ray telescopes, are found to have strong applicability to low energy nuclear physics. Their excellent timing resolution makes them very suitable for fast timing measurements and their much improved energy resolution compared to conventional scintillators promises to open up new avenues in nuclear physics research which were presently hard to access. Such "medium-resolution" spectroscopy has broad interest across several areas of contemporary interest such as the study of nuclear giant resonances. In addition to the connections to space science, it is striking that the demands of contemporary medical imaging have strong overlap with those of experimental nuclear physics. An example is the interest in PET-MRI combined imaging which requires putting scintillator detectors in a high magnetic field environment. This has led to strong advances in the area of silicon photomultipliers, a solid-state replacement for photomultiplier tubes, which are insensitive to magnetic fields. Broad application to nuclear physics of this technology may be foreseen.

1. Introduction

Gamma-ray spectroscopy is a key tool in experimental nuclear physics. It has many applications within the field, in particular, it is used to observe transitions between bound nuclear excited states. The observation and characterisation of such transitions in terms of their strength is an important test of contemporary models of nuclear structure. Such transitions between bound states are typically in the energy range 100 keV to a few MeV depending on the mass of the nucleus of interest, while giant resonances are also often probed through their gamma-ray emission and, here, the energy range of interest extends up to 10s of MeV. In all applications, a key driver is energy resolution. Over the last thirty years or so, the high-purity germanium detector has achieved particular prominence given its excellent energy resolution of 2-3 keV at 1.33 MeV. Given that in many areas of interest, there is a high multiplicity of emitted gamma rays leading to a highly complex spectrum, detectors are often deployed in large, multi-element arrays. Traditionally, Compton suppression was employed to improve the peak-to-total but in the last decade, gamma-ray tracking arrays of pure germanium have been devloped which can reconstruct the position of individual gamma-ray interactions to high precision using pulse shape analysis. With the advent of germanium detectors, the well-established technology of scintillator detectors has taken a back seat as far as gamma-ray spectroscopy has been concerned. It has

Journal of Physics: Conference Series **620** (2015) 012001 doi:10.1088/1742-6596/620/1/012001

been largely relegated to the study of giant resonances, where good efficiency for high-energy gamma rays is important but in general, good energy resolution is not warranted. The other main area where scintillator detectors are used extensively is in fast-timing measurements, which allow the lifetimes of excited states of the order of 100 ps to be measured. Historically, these were carried out with barium fluoride detectors, using the fast component of the BaF_2 signal to extract the timing information.

2. Next-generation scintillators in nuclear physics

Next-generation inorganic scintillator materials such as lanthanum bromide, cerium bromide and related materials have reawakened the relevance of scintillator detectors in low energy nuclear physics. Such detectors may achieve energy resolutions of around 3% at 662 keV which is a factor of two better than standard scintillator materials such as NaI(Tl) and CsI(Tl). They also have excellent timing resolution and relatively high intrinsic efficiency. The excellent timing resolution of next-generation scintillators like $LaBr_3$:Ce has allowed them to be rapidly applied to fast-timing measurements, displacing the earlier technology of BaF_2 detectors. The key advantage here is their energy resolution, which makes it easier to select γ-ray transitions of interest and study complex decay schemes [1, 2]. Fast-timing detectors are generally small, cylindrical and relatively cheap which has led to a quick uptake of the next-generation technology.

The improved energy resolution of next-generation scintillators could also be applied to what one might term "medium-resolution" spectroscopy. This is applicable to discerning structure within giant resonances e.g. the so-called pygmy dipole resonance [3]. It would also be of high relevance to investigation of exotic nuclear phenomena such as Jacobi shape transitions [4], and the study of light nuclei where transition energies are often large (few MeV) and the level density is relatively low.

The main barrier to a wide application of next-generation scintillators for medium-resolution spectroscopy is the high intrinsic cost of the material. The focus has been on $LaBr_3$ but some groups have obtained individual $CeBr_3$ detectors which are somewhat cheaper but have slightly poorer energy resolution. Other detector materials such as SrI_2 and CLYC have been investigated by nuclear physicists but their high cost and other factors have precluded their extensive use thus far. The high cost of the materials is compounded by the desire to produce close-packed arrays of detectors which needs detectors with a square geometry. Different approaches have been taken to this problem. For example, the PARIS calorimeter [5], foreseen to ultimately comprise an array of 200 detectors, will comprise two shells of detectors: an inner shell of $LaBr_3$(Ce) and an outer shell of NaI(Tl). The inner shell will have high resolution up to 10 MeV while the outer shell is designed to efficiently stop gamma rays up to 50 MeV. The present design of PARIS comprises individual phoswich elements [6], each of a 2" cubic $LaBr_3$(Ce) crystal coupled to a 2" x 2" x 6" NaI(Tl) crystal, with the scintillation light collected with a common photomultiplier tube. Pulse shape analysis may be used to disentangle the signals from the two crystal elements based on their significantly different rise and decay times (see Fig. 1). This is challenging in terms of digital electronics. The high intrinsic activity of $LaBr_3$(Ce) may prove challenging when a large calorimeter is assembled, although in general most measurements would be made in coincidence with detection of another radiation type such as charged particles or in coincidence with a pulsed beam, which could remove random coincidences with intrinsic background.

An example application of PARIS may be to the study of alpha clustering in nuclei, for example, by studying the very weak decay branches between highly-excited cluster states [7]. Fig 2 compares real data obtained on the $^{12}C(^{12}C,\gamma)$ reaction using a BGO scintillator array and the DRAGON spectrometer at TRIUMF [8] with the results of a GEANT3 simulation of the same process using the PARIS calorimeter. The significant advantages of the $LaBr_3$(Ce) component in resolving individual excited states in this reaction are very evident.

It is interesting to note, significant advances in imaging with large $LaBr_3$ crystals. It is not

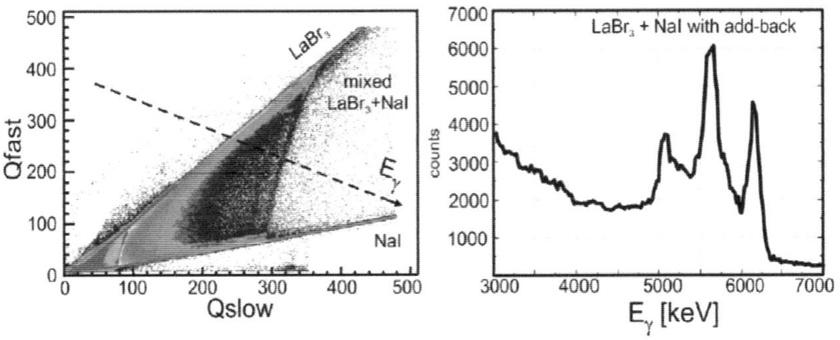

Figure 1. Response of PARIS phoswich detector to 6.13 MeV gamma rays. (Left) Two-dimensional plot of fast and slow components of the signal originating from $LaBr_3$ and NaI(Tl) respectively (Right) The reconstructed energy spectrum following add-back of the two signals [10]

obvious with large crystals that the position where scintillation light is recorded retains any memory of where it is produced in the crystal. However, Giaz *et al.* [9] have recently shown that 1-cm position resolution can be achieved in a 3" cubic $LaBr_3$ crystal using a segmented photomultiplier tube and a collimated source. This discovery is of some signficance for nuclear physics as Doppler correction of gamma rays must be applied and so knowing the angle of interaction well helps with this. Moreover, Doppler broadening is reduced if the effective opening angle of the detector is constrained.

3. Silicon photomultipliers

Silicon photomultipliers (SiPMs), which are dense arrrays of avalanche photodiodes (APDs), are a disruptive technology that may, in the coming years, displace conventional photomultipliers for many applications. Their performance is well-matched to the requirements of scintillator detectors in nuclear physics since they can preserve good energy resolution and timing resolution. In particular, they are insensitive to magnetic fields, unlike conventional photomultipliers. A major driver in the development of SiPMs has been the PET imaging sector. Indeed, there is currently a push to integrate PET with MRI (magnet resonance imaging) [11, 12] which emphasises the need for photosensors which are insensitive to magnetic fields. The medical advantages in this area are clear since functional and anatomical information can be obtained simultaneously in a single image, providing improved diagnostics leading to a more effective treatment. The main difficulty in achieving this is that PMTs are affected by magnetic fields and it is impractical to shield them within the high field of an MRI magnet. Practically speaking, there are various ways to achieve combined PET-MRI: for example, splitting the magnetic field volume so that the PMTs sit in zero field, or using fibre-optic cables (a nonmagnetic component) to transfer light from the scintillation crystals to PMTs which are placed outside of the magnetic field [13]. The other approach to PET-MRI is to use APDs or SiPMs which can be placed directly in the magnetic field [14, 15, 12]. Several different prototype systems have been successfully demonstrated. For example, Woody *et al.* explored their RatCAP detector containing 12 LSO (lutetium oxyorthosilicate) arrays coupled to APDs and found that it was not affected by magnetic fields and strong RF pulses, allowing them to succeed in obtaining a high quality medical image [15].

There are also a number of areas of nuclear physics where gamma-ray detectors could be operated in regions of high magnetic field, for example in the vicinity of large spectrometers.

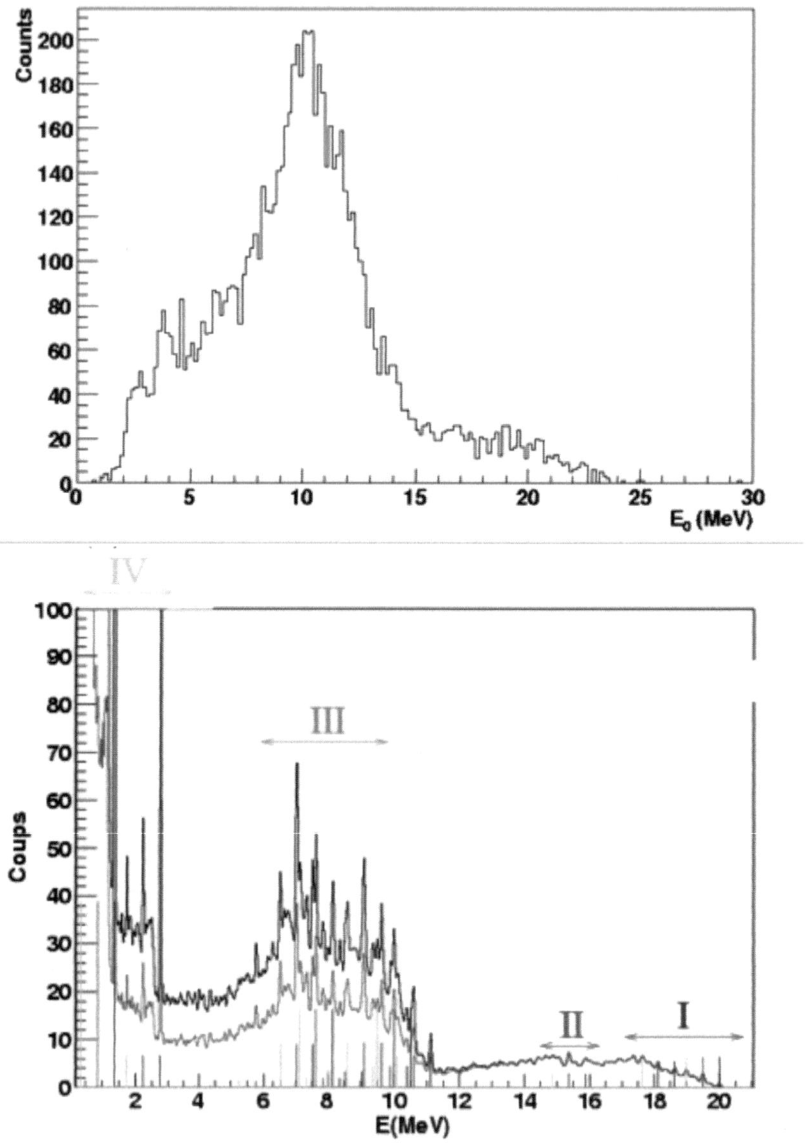

Figure 2. (top) Gamma-ray energy spectrum for the $^{12}C(^{12}C,\gamma)$ reaction at $E_{c.m}$=6.0 MeV measured with a BGO detector array (bottom) Simulation of the same experiment using the PARIS calorimeter which incorporates an inner shell of $LaBr_3$ detectors.

An example where detectors would be placed directly within a high field is the HELIcal Orbit Spectrometer (HELIOS) built at Argonne National Laboratory [16] which is used to study single-particle transfer reactions in inverse kinematics. The basic component of this device is a large-bore solenoidal magnet, in fact, a former hospital MRI magnet. The addition of scintillator-based gamma-ray detectors inside the field volume would present advantages in understanding nuclear structure through decay cascades, and if timing resolution were sufficiently high then lifetimes of states could be obtained in some cases and decay cascades ordered. Operating γ-ray detectors inside such a solenoidal spectrometer would have all the same demands and constraints as for PET-MRI, and SiPM-based readout is again an attractive solution. Such a system could be envisaged for a solenoidal spectrometer presently under construction for the HIE-ISOLDE facility at CERN [17]. The SiPM technology is already moving very rapidly. Fig. 3 shows an 8 x 8 array of 6-mm square C-series SiPMs available from SensL. This sensor is well-matched to a 2" cubic scintillator crystal and the multi-channel output would make it suitable for exploring the imaging applications discussed above [9].

Figure 3. An 8X8 array of 6-mm SiPMs from SensL. The array is mounted on a inspection board with the individual energy and fast-timing outputs coupled together.

4. Conclusions

Next-generation scintillator detectors offer considerable potential for employment in low energy nuclear physics on the basis of their excellent energy and timing resolution. Such detectors have rapidly been adopted in the area of fast timing, superseding the earlier approach using BaF_2 detectors. There is strong interest in using larger detectors of next-generation scintillators for medium-resolution spectroscopy, opening up new avenues of research in the areas of giant resonances and exotic nuclear shapes. Take-up of technology here has been slower due to the high intrinsic cost of the material. Nevertheless, initiatives such as the PARIS calorimeter are already beginning to explore the potential in this area. Silicon photomultipliers are another new technology which is advancing very quickly in performance, driven by demands of medical imaging and other areas. This also has strong application to nuclear physics, both to replace PMTs in conventional applications but also to open up scope to readily put high-performing gamma-ray detectors directly in high magnetic field environments. The drive by nuclear physicists to obtain best possible energy resolution and performance from scintillators and photosensors places them in an enviable position for exploiting this technology for societal applications in homeland security and other areas.

References

[1] Deloncle I *et al* 2010 *J. Phys. Conf. Ser.* **205** 012044
[2] Roberts O J *et al* 2013 *J. Phys. Conf. Ser.* **381** 012124
[3] Ponomarev V 2014 *J. Phys. Conf. Ser.* **533** 012028
[4] Ward D *et al* 2002 *Phys. Rev. C* **66** 024317
[5] Maj A *et al* 2009 *Acta Physica Polonica B* **40** 565
[6] Roberts O J *et al* 2011 *Acta Physica Polonica B* **42** 729
[7] Jenkins D 2013 *Electromagnetic transitions as a probe of clustering in nuclei* (Clusters in Nuclei vol III) ed C Beck *Lecture Notes in Physics* **875** pp 25-49
[8] Jenkins D G *et al* 2007 *Phys. Rev. C* **76** 044310
[9] Giaz A *et al* 2015 *Nucl. Instrum. Meth. A* **772** 103
[10] Zieblinski M *et al* 2013 *Acta Phys. Pol. B* **44** 651
[11] Judenhofer M S *et al* 2008 *Nature Medicine* **14** 459
[12] Catana C *et al* 2008 *PNAS* **105(10)** 3705-3710
[13] Shao Y *et al* 1997 *Physics in Medicine and Biology* **42** 1965-1970
[14] Hong K J *et al* 2013 *Med. Phys.* **40** 042503
[15] Woody C *et al* 2007 *Nucl.Instrum. Meth. A* **571** 102-105
[16] Wuosmaa A H *et al* 2007 *Nucl. Instrum. Meth. A* **580** 1290
[17] Jenkins D G 2014 *Nature Physics* **10** 909

Development of glass-ceramic scintillators for gamma-ray astronomy

D de Faoite[1], L Hanlon[2], O Roberts[2], A Ulyanov[2], S McBreen[2], I Tobin[2] and K T Stanton[1]

[1] UCD School of Mechanical and Materials Engineering, University College Dublin, Belfield, Dublin 4, Ireland

[2] UCD School of Physics, University College Dublin, Belfield, Dublin 4, Ireland

E-mail: daithi.defaoite@ucd.ie

Abstract. Scintillators synthesised as glass-ceramics have several potential benefits compared to the currently-used halide scintillators, including non-hygroscopicity, mechanical ruggedness, ease of producing customisable shapes, and the potential for low-cost synthesis. The use of these scintillators is considered for a gamma-ray telescope operating in the 0.2 MeV–50 MeV photon range. Inorganic scintillator compounds suitable for incorporation into glass-ceramics are assessed. In addition, several families of glass suitable for use as hosts for scintillating compounds are also reviewed.

1. Introduction

Scintillation detectors have been used in gamma-ray astronomy telescopes such as the Fermi Gamma-ray Space Telescope and the Compton Gamma Ray Observatory. The requirement to improve the sensitivity of future gamma-ray telescopes, particularly in the 0.2 MeV–10 MeV photon range has motivated an analysis of different scintillator options.

The most commonly used material types for gamma-ray spectroscopy applications are inorganic single crystals, transparent polycrystalline ceramics, glasses, polymers loaded with scintillating compounds, and inert gases, each having advantages and disadvantages [1]. Families of inorganic materials widely-used as scintillators include the alkali, alkaline earth, and rare earth halides, often doped with rare earth elements. These families of materials, which include the most widely used inorganic scintillator NaI(Tl), contain compounds that exhibit high light yield and good timing and energy resolutions. However, they are hygroscopic, and must therefore be hermetically encapsulated. The halide scintillators are normally used in monocrystalline form, and are typically grown by either the Czochralski technique [2, 3], micro-pulling-down technique [4, 5], or by using a Bridgman furnace [6]. The cost of growing single-crystals using these techniques is very high, and can be many times the cost of the raw materials [7].

Transparent ceramic scintillators are produced in polycrystalline form by high temperature sintering of compacts of nano-sized powders, generally followed by hot pressing. While somewhat less expensive than monocrystalline crystal growth methods, production of scintillators by sintering and hot pressing still results in high scintillator costs. Additionally, the hot pressing method is normally limited to production of simple cylindrical shapes. The polycrystalline ceramics produced are generally non-hygroscopic and significantly more rugged than halide scintillators. Another class of inorganic scintillators are scintillating glasses. ^{6}Li-doped glasses are widely used for thermal neutron detection and spectroscopy.

Nanocrystalline glass-ceramics are a promising new class of inorganic scintillator. Kang *et al.* and Barta *et al.* demonstrated the feasibility of using glass-ceramic scintillators for gamma-ray spectroscopy [8, 9]. Glass-ceramics possess several advantages over halide and ceramic scintillators.

1.1. Scintillator Requirements

For the current application a scintillator with an energy resolution of less than 5 % at 662 keV is required. To achieve this the scintillator must have a high light yield and good light yield linearity. A peak luminescent emission well-matched to the peak sensitivity of silicon photomultiplier detectors (SiPM) is also required. SiPMs are the likely photodetector that will be used for a gamma-ray telescope. A short primary decay constant of < 100 ns is required to make the detector suitable for high count rate applications. To make low count rate experiments feasible a scintillator with low intrinsic activity is required. Scintillators containing significant quantities of long-lived meta-stable radioisotopes such as ^{138}La, ^{176}Lu or ^{87}Rb are therefore unsuitable for the current application. A high density ($> 5\,\mathrm{g \cdot cm^{-3}}$) and high effective atomic number is required to provide a sufficiently high gamma-ray attenuation coefficient, and to favour photoelectric interaction over Compton scattering. The material should not be hygroscopic as this would require the crystal to be encapsulated, resulting in significant dead space in the telescope. The scintillator material should be available in a wide range of sizes, and be physically robust. Because a large quantiy of scintillator material will be required, the specific cost of the scintillator should be moderate. The current cost of producing monocrystalline $LaBr_3$(Ce) and $CeBr_3$ scintillators makes their use prohibitively expensive at the proposed instrument scale.

2. Glass-Ceramics

Glass-ceramics are engineered materials formed by the controlled nucleation and crystallisation of glass. Glass-ceramics are commonly produced by a melt-quench process, followed by annealing, but may be produced by other methods such as the sol-gel process [10]. Controlled crystallisation is achieved by a two-stage heat-treatment process, whereby crystals nucleate and then grow *in situ* from the amorphous glass. As a consequence of this, crystallite distribution can be uniform throughout the glass body. The nature of glass-ceramics allows numerous optical and mechanical properties to be controlled by suitable adjustment of the batch composition and heat treatment regime. Glass-ceramics are widely used for their controllable properties, combining the desirable characteristics of glasses and sintered ceramics.

Unlike crystalline solids, glasses are not thermodynamically stable. Their internal energy is slightly higher than that of a crystalline material of the same composition. A thermodynamic driving force therefore exists for the re-arrangement of the atoms in a glass to form crystalline phases. However, if a material is cooled sufficiently rapidly from the molten state, there will be insufficient time available for atoms to re-arrange themselves to form crystalline phases. Once the material has sufficiently cooled, the atoms possess insufficient thermal energy to rearrange themselves into an ordered lattice, and they will be trapped in the amorphous state. Due to their thermodynamically unstable nature, many glasses may be crystallised in a controlled manner by suitable heat treatment.

The composition of a parent glass may be chosen so that the crystalline phases that are formed upon heat treatment are scintillating compounds. A material so-designed is termed a glass-ceramic scintillator (GCS). The feasibility of using glass-ceramic scintillators for gamma-ray spectroscopy and radiation detection has already been demonstrated by Barta *et al.* and Kang *et al.*, although the energy resolution so far obtained has been low [9, 8, 11, 12].

Glass ceramics have several properties that make them desirable candidate materials for gamma-ray scintillator applications. They are more mechanically rugged and are simpler to synthesise in large volumes than halide scintillators. Glass-ceramics do not cleave, and may be cut and polished with relative ease. The glass matrix of a GCS provides environmental protection to the scintillating compounds, allowing hygroscopic scintillating compounds to be used without encapsulation, while the crystallites act as the luminescent centres. The glass-ceramic production route allows for more compositional flexibility than is possible for monocrystalline crystal growth methods, and may be used to produce crystallites of compounds that do not melt congruently. More significantly, the cost of producing glass-ceramics is typically far less than the crystal growth methods used for halide scintillators, or the hot pressing method used for polycrystalline ceramics, providing the potential to significantly reduce the cost of producing inorganic scintillators.

It is desirable to maximise the volume fraction of scintillating phase crystallites in a glass-ceramic in order to maximise the visible light output. However, excessively high crystallite volume fractions may lead to loss of optical transmissivity [9]. The compositions and heat treatment regimes of GCSs must therefore be carefully optimised to achieve a suitable balance of scintillation and other optical and physical properties. Optical transparency in a glass-ceramic containing a moderately high crystalline volume fraction may be retained by forming crystallites with diameters on the order of 20 nm–50 nm or less [13, 14]. Adjustment of the heat treatment parameters allows the size of the crystallites to be controlled. Matching of the refractive indices of the base glass and crystallites also minimises light scattering.

Polymers loaded with nano-sized scintillating crystallites are also under investigation for scintillation applications. However, glass-matrix nano-structured materials appear more promising as they have higher gamma-ray attenuation than polymer-matrix scintillators [13]. In addition, because the scintillating crystallites in GCSs are formed *in situ*, agglomeration problems that can occur with loaded polymer-matrix scintillators are avoided [13].

Glass-ceramics are preferred to glass for scintillator applications due to the potentially higher light yields obtainable. Glass scintillators exhibit low transfer efficiency of excitation energy from the matrix to the luminescent centres, caused by the abundance of charge carrier trapping sites in glass matrices [15, 14]. These charge carrier traps, which originate from network defects such as non-bridging oxygen and impurities, can lead to non-radiative recombination (phonon emission), and can thereby result in poor light yield. The ordered nature of the crystalline phases in glass-ceramics may lead to more efficient energy transfer to the luminescent centres and a higher probability of radiative recombination [16].

2.1. Glass Systems for Glass-Ceramic Scintillators

Several types of parent glass may be used as a basis for glass-ceramic scintillators. The parent glass should ideally provide a matrix that is capable of incorporating a large volume fraction of scintillating phase in order to maximise light yield. It should provide sufficient chemical stability to protect hygroscopic scintillating compounds from environmental moisture. In addition, the glass matrix should provide good mechanical strength, and should be optically transparent.

Silicate or aluminosilicate glasses are suitable candidate glass families for synthesis of glass-ceramics. Fluxes such as Na_2O and CaO are commonly added to silicate glass to reduce the required processing temperature. These fluxes reduce the connectivity of the glass network, thereby reducing the viscosity of the glass. Some rare earth dopant ions tend to form clusters in silicate glasses, leading to premature concentration quenching [17, 9]. Addition of aluminium cations can prevent rare earth ion clustering, thereby enhancing luminescent efficiency [18, 9].

Oxyfluoride glasses are also suitable parent glass systems for synthesis of glass-ceramic scintillators. Oxyfluoride glasses provide a low phonon energy fluoride environment suitable for rare earth ion doping, while retaining the desirable mechanical and chemical stability of

Journal of Physics: Conference Series **620** (2015) 012002 doi:10.1088/1742-6596/620/1/012002

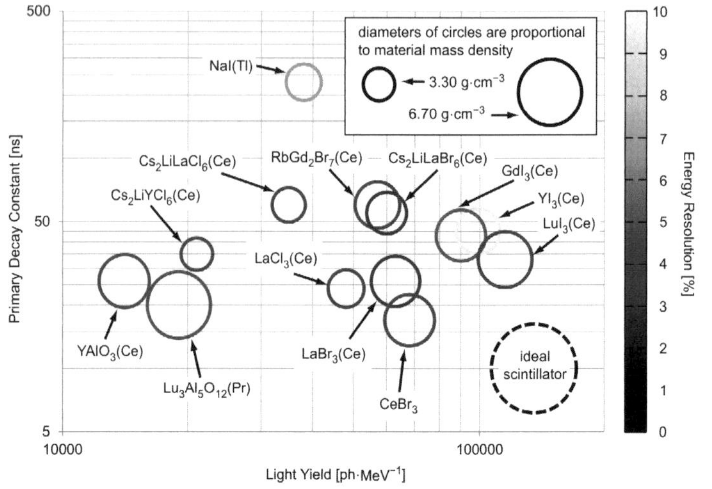

Figure 1. The light yield, primary decay constant, density, and energy resolution of several scintillators.

oxide glasses [19]. Nano-sized crystallites may be precipitated in many oxyfluoride glasses due to the formation of a diffusion barrier surrounding the crystals [20, 21, 22]. In addition to scintillator applications, oxyfluoride glass-ceramics doped with rare earth ions are also of interest for upconversion lasers and telecommunications optical amplifiers [19].

3. Scintillator Compounds

Rare earth halide scintillating compounds such as $LaBr_3$(Ce) and $CeBr_3$ have many desirable properties, including excellent timing and energy resolutions. However, the high cost, hygroscopicity, and difficulty of manufacturing large single crystals of these materials are undesirable. Numerous ceramic scintillators have also been investigated for scintillator applications including aluminium perovskites, aluminium garnets, oxyorthosilicates, and pyrosilicates. Several of these scintillators also exhibit desirable gamma-ray scintillation properties, combined with ruggedness and non-hygroscopicity. However, the energy resolutions so-far demonstrated by ceramic scintillators have been inferior to those of the best rare earth halides. Additionally, despite being less expensive than monocrystalline growth methods, the sintering and hot pressing route used to produce these transparent ceramics still results in rather high scintillator costs.

The light yield, primary decay constant, density, and energy resolution of several promising scintillator compounds suitable for integration into glass-ceramics are illustrated in Figure 1. Several scintillators with high intrinsic activities, which would not be suitable for the current application, are included in Figure 1 for comparison purposes.

4. Glass-Ceramic Scintillators

To-date, a limited number of the known inorganic scintillating compounds have been produced as glass-ceramics. Several of these are listed in Table 1, along with the glass batch composition used. Many of these glass-ceramics have been investigated for applications other than gamma-ray spectroscopy, including photonic and laser upconversion applications. For this reason the scintillation properties of many of these glass-ceramics have not been characterised.

Fu *et al.* produced transparent CaF_2 glass-ceramics doped with 0.1 mol% Eu^{2+} ions [19]. CaF_2 crystallites in the size range 11 nm–18 nm were precipitated from the glass by heat treating at

Table 1. Summary of some known inorganic scintillator compounds that have been produced as glass-ceramics.

Compound	Glass Family	Batch Composition	Reference
BaF_2	oxyfluoride	$(100-x)(2Na_2O-16K_2O-8Al_2O_3-74SiO_2){\cdot}xBaF_2$	[22]
$Bi_4Ge_3O_{12}$	bismuthate-germanate	$40Bi_2O_3-60GeO_2$	[23]
CaF_2(Eu)	oxyfluoride	$45SiO_2-20Al_2O_3-10CaO-25CaF_2$	[19]
CaF_2(Eu)	oxyfluoride	$49\,SiO_2-20\,Al_2O_3-30\,CaF_2-1\,EuF_2$	[24]
$CeBr_3$	oxyfluoride	$50\,SiO_2-16\,Al_2O_3-19\,NaF-13\,GdBr_3-2\,CeBr_3$	[8]
CeF_3	oxyfluoride	$40\,SiO_2-28\,Al_2O_3-17\,NaF-15\,CeF_3$	[25]
CeF_3	fluorophosphate	$NaPO_3-MF_2-CeF_3$, M = Pb, Zn, Cd, Ba	[26]
$GdBr_3$(Ce)	oxyfluoride	$50\,SiO_2-16\,Al_2O_3-19\,NaF-13\,GdBr_3-2\,CeBr_3$	[8]

700 °C. The luminescence intensity of the glass-ceramic was found to be more than twice that of the base glass, caused by the migration of Eu^{2+} dopant ions into the CaF_2 crystallites. Precipitation of crystallites slightly shifted the optical absorption edge to longer wavelength, but did not significantly degrade the transparency of the glass.

Secu *et al.* produced CaF_2(Eu) glass-ceramic in an oxyfluoride glass matrix [24]. Heat treatment at 760 °C for 15 min led to the formation of CaF_2(Eu) nano-crystals with a mean size of 65 nm. Photoluminescence emission for the glass-ceramic was found to peak at 425 nm, ascribed to the Eu^{2+} f–d transition [24]. Neither Fu *et al.* nor Secu *et al.* characterised the scintillation performance of their CaF_2(Eu) glass-ceramics.

Chen *et al.* synthesised a glass-ceramic containing CeF_3 nano-crystals using an oxyfluoride glass matrix [25]. Crystallites were precipitated by heat treating at 630 °C for 2 h. The morphology of the resulting glass-ceramic was a monodisperse arrangement of spherical CeF_3 nano-crystals with diameters of 10 nm–15 nm. The glass-ceramic exhibited a broad luminescent emission spectrum peaking at 407 nm, with a decay constant of 42 ns. Chen *et al.* investigated the glass-ceramic for light emitting diode applications rather than for scintillating applications, and therefore did not characterise the scintillation properties of this CeF_3 glass-ceramic.

Jiang *et al.* investigated ternary phosphate glasses containing cerium for scintillator applications [26]. Of the glass systems investigated, $NaPO_3-CeF_3-BaF_2$ allowed the highest concentration of cerium to be incorporated into the glass. Jiang *et al.* found that the scintillation light output of this glass increased with increasing cerium concentration, and reached 60 % of that of crystalline CeF_3 with a CeF_3 concentration of 25 mol%.

Kang *et al.* investigated the scintillation performance of $GdBr_3$(Ce) in an oxyfluoride glass matrix for gamma-ray spectroscopy applications. A full energy photopeak was recorded for the material when exposed to a ^{137}Cs source, and an energy resolution of 27 % was obtained for 662 keV gamma-rays. The light yield of the $GdBr_3$(Ce) activated glass was found to be 1700 ph $\cdot$ MeV^{-1}, and the primary decay constant was found to be approximately 55 ns, attributed to Ce^{3+} emission. In a separate study Kang *et al.* report an improved light yield of approximately 2000 ph $\cdot$ MeV^{-1}–3000 ph $\cdot$ MeV^{-1} for a $GdBr_3$(Ce) glass-ceramic [13].

5. Discussion and Conclusions

5.1. Scintillator Compounds for Synthesis of Glass-Ceramics

Promising candidate scintillator compounds for integration into glass-ceramics include: YI_3(Ce), $LaCl_3$(Ce), $LaBr_3$(Ce), $CeBr_3$, GdI_3(Ce), and LuI_3(Ce). The high melting temperatures of ceramic scintillating compounds such as $YAlO_3$(Ce) makes production of glass-ceramics containing these compounds more challenging.

5.2. Potential of Glass-Ceramic Scintillators

Glass-ceramic scintillators are a promising new class of inorganic scintillator that offer many potential benefits [27, 8]. Glass-ceramic scintillators are non-hygroscopic, mechanically rugged, and are simpler to synthesise in large volumes than monocrystalline halide scintillators. The ruggedness of glass-ceramic scintillators makes them attractive not only for the current application, but also for applications in harsh environments such as oil-well logging. In contrast to nano-structured polymer matrix composites, crystallite distribution can be uniform throughout a glass-ceramic. Most significantly, the cost of producing glass-ceramics is typically far less than the crystal growth methods used for monocrystalline halide compounds, or the hot pressing method used for polycrystalline ceramics, providing the potential to significantly reduce the cost of producing inorganic scintillators.

The main disadvantage of GCSs to-date has been the poor light yields and energy resolutions obtained. However, careful optimisation of the compositions and heat treatment regimes of glass-ceramic scintillators may yield improvements in light yield and energy resolution.

Acknowledgments

The authors would like to acknowledge funding from the European Space Agency.

References

[1] Greskovich C and Duclos S 1997 *Annual Review of Materials Science* **27** 69–88
[2] Czochralski J 1918 *Zeitschrift für Physikalische Chemie* **92** 219–221
[3] Nikl M, Yoshikawa A, Vedda A and Fukuda T 2006 *Journal of Crystal Growth* **292** 416–421
[4] Yoon D H, Yonenaga I, Fukuda T and Ohnishi N 1994 *Journal of Crystal Growth* **142** 339–343
[5] Fukuda T, Rudolph P and Uda S (eds) 2004 *Fiber Crystal Growth from the Melt* (Springer)
[6] Zaslavsky B G 1999 *Journal of Crystal Growth* **200** 476–482
[7] Derenzo S E, Weber M J, Bourret-Courchesne E and Klintenberg M K 2003 *Nuclear Instruments and Methods in Physics Research Section A* **505** 111–117
[8] Kang Z T, Rosson R, Barta B, Han C, Nadler J H, Dorn M, Wagner B and Kahn B 2013 *Radiation Measurements* **48** 7–11
[9] Barta M B 2012-12 *Nanocomposite Glass-Ceramic Scintillators for Radiation Spectroscopy* Master's thesis
[10] Biswas A, Maciel G S, Friend C S and Prasad P N 2003 *Journal of Non-Crystalline Solids* **316** 393–397
[11] Barta M B, Nadler J H, Kang Z, Wagner B K, Rosson R, Cai Y, Sandhage K H and Kahn B 2014 *Applied Optics* **53** D21–D28
[12] Kang Z, Wagner B K, Nadler J H, Rosson R, Kahn B and Barta M B 2014-06-19 Transparent glass scintillators, methods of making same and devices using same
[13] Kang Z, Wagner B K, Summers C J, Nadler J, Rosson R and Kahn B 2012-10-27/2012-11-03 *IEEE Nuclear Science Symposium and Medical Imaging Conference (NSS/MIC)* pp 1688–1691
[14] Barta M B, Nadler J H, Kang Z, Wagner B K, Rosson R and Kahn B 2013 *Optical Materials* **36** 287–293
[15] Baccaro S, Cecilia A, Mihokova E, Nikl M, Nitsch K, Polato P, Zanella G and Zannoni R 2002 *Nuclear Instruments and Methods in Physics Research Section A* **476** 785–789
[16] Kang Z, Wagner B K, Nadler J H, Rosson R, Kahn B and Barta M B 2013-02-14 Transparent glass scintillators, methods of making same and devices using same
[17] Ishizaka T, Kurokawa Y, Makino T and Segawa Y 2001 *Optical Materials* **15** 293–299
[18] Arai K, Namikawa H, Kumata K, Honda T, Ishii Y and Handa T 1986 *Journal of Applied Physics* **59** 3430
[19] Fu J, Parker J M, Flower P S and Brown R M 2002 *Materials Research Bulletin* **37** 1843–1849
[20] Bhattacharyya S, Bocker C, Heil T, Jinschek J R, Höche T, Rüssel C and Kohl H 2009 *Nano Letters* **9** 2493–2496
[21] Bocker C, Bhattacharyya S, Höche T and Rüssel C 2009 *Acta Materialia* **57** 5956–5963
[22] Bocker C, Avramov I and Rüssel C 2010 *Chemical Physics* **369** 96–100
[23] Polosan S, Galca A C and Secu M 2011 *Solid State Sciences* **13** 49–53
[24] Secu M, Secu C E, Polosan S, Aldica G and Ghica C 2009 *Journal of Non-Crystalline Solids* **355** 1869–1872
[25] Chen D, Yu Y, Lin H, Huang P, Weng F, Shan Z and Wang Y 2009 *Optics Letters* **34** 2882–2884
[26] Jiang J, Zhang G and Poulain M 1998 *Journal of Alloys and Compounds* **275-277** 733–737
[27] Wagner B K, Kang Z, Nadler J, Rosson R and Kahn B 2012-05-01 *Proceedings of SPIE 8373, Micro- and Nanotechnology Sensors, Systems, and Applications IV* **8373** 83730K

Characterization of new scintillators: SrI_2:Eu, $CeBr_3$, GYGAG:Ce and CLYC:Ce

A. Giaz[1], V. Fossati[2], G. Hull[3], F. Camera[1-2], N. Blasi[1], S. Brambilla[1], S. Ceruti[1-2], N. Cherepy[4], B. Million[1], L. Pellegri[1] and S. Riboldi[1-2]

[1]Istituto Nazionale di Fisica Nucleare INFN, Sezione di Milano, Via Celoria 16, 20133 Milano (Mi), Italy.
[2]Università degli studi di Milano, Via Celoria 16, 20133 Milano (Mi), Italy.
[3]Institut de Physique Nucleaire d'Orsay, 15 rue Georges Clemenceau, 91406 ORSAY CEDEX.
[4]Lawrence Livermore National Laboratory, Livermore, CA 94550.

E-mail: agnese.giaz@mi.infn.it

Abstract.
We measured the performance of a 2"x2" tapered cylindrical SrI_2:Eu, a 2" x 3" cylindrical $CeBr_3$, a 2" x 0.3" cylindrical GYGAG:Ce. The gamma-ray energy resolution was measured up to 9 MeV. A scan along three axes of each scintillator was performed using a collimated source of ^{137}Cs. The signals of the detectors were also digitized and compared. We tested two 1" x 1" cylindrical CLYC crystals. One was enriched with ^{6}Li at 95% while the other was enriched with 99% of ^{7}Li. The response of the two detectors to gamma rays, thermal and fast neutrons was measured. The PSD performances have been tested with a different types of PMTs.

1. Introduction

There is an intense R&D work on $CeBr_3$ and SrI_2:Eu crystals as larger crystals are becoming available [1-2]. The development of new ceramic scintillator materials (such as GYGAG:Ce) offers the possibility to have high-resolution gamma-ray spectroscopy at low cost [3]. The scintillator Cs_2LiYCl_6:Ce (CLYC) is a very interesting material for radiation detection because of its ability to measure gamma rays and neutrons simultaneously [4]. CLYC scintillators are suitable for thermal neutrons detection, owing to ^{6}Li content and they can also be used as fast neutron spectrometers, owing to ^{35}Cl content. The gamma rays and neutrons can be discriminated by pulse shape discrimination methods [5].

In this work the results of the characterization measurements of SrI_2:Eu, $CeBr_3$ and GYGAG:Ce are described in section 2. The measurements performed with two samples of CLYC scintillators, characterized by different Li ions concentrations, are discussed in section 3; in particular, the general properties, the fast and thermal neutron detection capability, and the PSD performances with different PMTs are presented in section 3.1, 3.2 and 3.3, respectively. The conclusions are given in section 4.

2. Characterization of SrI_2:Eu, $CeBr_3$ and GYGAG scintillators sample

The SrI_2:Eu and the GYGAG:Ce samples have been provided by the Lawrence Livermore National Laboratory while the $CeBr_3$ crystal by the Institut de Physique Nucleaire Orsay. The measurements were performed in the gamma spectroscopy laboratory of the University of Milan. The 2" x 2" cylindrical tapered SrI_2:Eu and the 2" x 3" cylindrical $CeBr_3$ were coupled to a HAMAMATSU R6233-100sel PMT powered at 800V and 850V, respectively. The 2" x 0.3" GYGAG:Ce was coupled to a HAMAMATSU R6231-100mod (HV=800V). The used Voltage Dividers (VD) were HAMAMATSU E1198-27 for all the detectors. Since $CeBr_3$ signal is very similar to that of $LaBr_3$:Ce, for this scintillator we used as well the LABRVD. This dedicated VD was developed by the electronic workgroup of the University of Milan for $LaBr_3$:Ce scintillators to reduce the PMT induced non-linearity at high energies [6].

The scintillation response was measured using standard gamma ray sources (^{22}Na, ^{60}Co, ^{88}Y, ^{133}Ba, ^{137}Cs, ^{152}Eu) and an AmBe(Ni) composite source. The signals of the three detectors were sent to a spectroscopic amplifier (TENNELEC TC244) and, in turn, to an ADC (ORTEC ASPEC MCA 926). The ^{152}Eu spectra, acquired with all tested scintillators, are compared in Fig. 1, while the AmBe(Ni) energy spectra are compared in Fig. 2. At 662 keV we measured a FWHM-energy resolution equal to 4.0%, 4.4% and 5.2% for the SrI_2:Eu, the $CeBr_3$ and the GYGAG:Ce respectively[7].

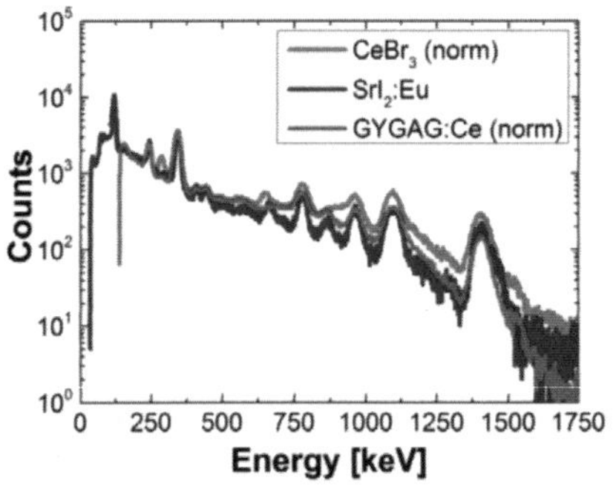

Figure 1. The ^{152}Eu energy spectra acquired using a standard spectroscopic amplifier for the three tested scintillators. The SrI_2:Eu has the best energy resolution among these scintillators. The spectra are normalized on the area of the 344 keV peak of the ^{152}Eu source.

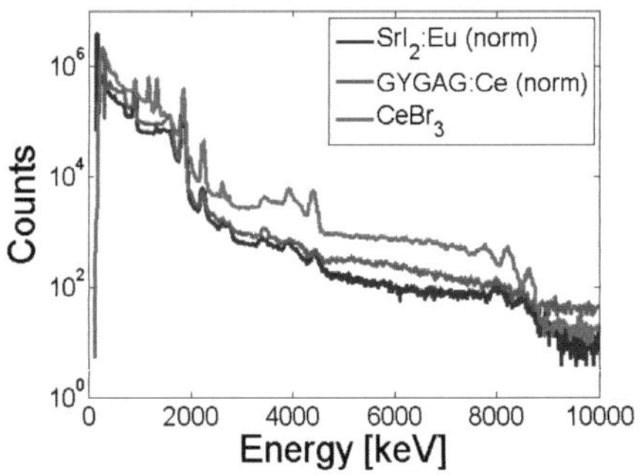

Figure 2. The measured SrI_2:Eu, GYGAG:Ce and $CeBr_3$ acquired using a PMT, a standard spectroscopic amplifier and an ADC. The used sources were AmBe(Ni), ^{88}Y and ^{60}Co (only for $CeBr_3$).

The detector anode pulses were also digitized using a 12 bit LeCroy HDO 6054 oscilloscope. A set of pulses (~1000) at a fixed energy were averaged to produce the reference signals. The signal properties (rise time and fall time) of all detectors were compared and the signals of each scintillator were studied from 662 keV up to 9 MeV. Fig. 3 shows the comparison of the three scintillator signal shapes, at 662 keV. The $CeBr_3$ provides the fastest signal: it has a rise time of ~18 ns and a fall time of ~70 ns. The SrI_2:Eu has the slowest signal among the tested detectors (fall time ~7 μs).

No significant change in the GYGAG:Ce and SrI_2:Eu signal shapes up to 9 MeV, whereas a small change was seen in $CeBr_3$ at 9 MeV of energy [7].

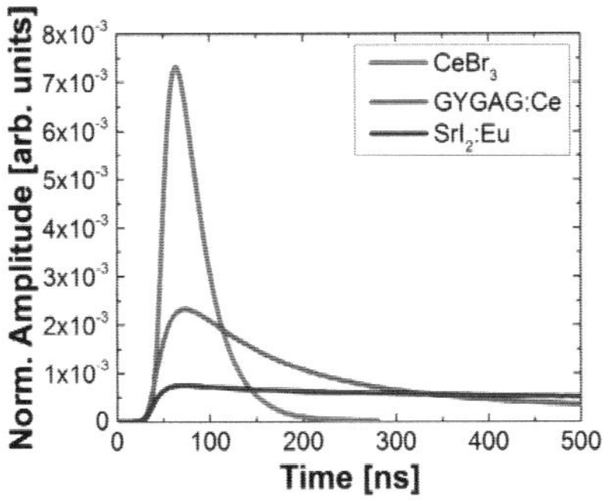

Figure 3. The normalized pulses of the three detectors for 662 keV gamma rays. The $CeBr_3$ has the fastest signal among these scintillators, whereas SrI_2:Eu has the slowest one.

The energy spectra and the pulses were also measured using a collimated beam of 662 keV gamma rays scanning the detector along the x, y, and z axes. A non-collimated ^{88}Y source, providing two calibration points, was placed nearby and used as a reference. It was possible to study how the position of the centroid, the FWHM, and the area of the 662 keV peak changes as a function of the position of the incident radiation. The detector response was measured to be constant for the $CeBr_3$ and the GYGAG:Ce. No significant variations of the peak position or the FWHM were observed. While SrI_2:Eu detector shows variations in the FWHM along the z axis imputable to the self-absorption effects, as shown in Fig. 4. The SrI_2:Eu provides the best energy resolution among the studied detectors (4.0 % at 662keV) but this value strongly depends on the position of the gamma-ray interaction point. A FWHM value as good as 22 keV at 662 keV (3.3%) was measured with the collimated source, as shown in Fig. 4 [7].

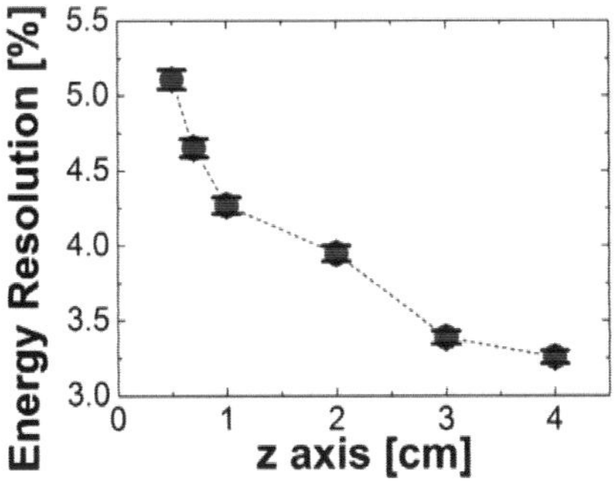

Figure 4. The FWHM of 662 keV peaks from a collimated ^{137}Cs source on the SrI_2:Eu crystal. The different values were obtained by moving the collimated beam along the z axis of the crystal. The spectra were calibrated using a non-collimated ^{88}Y sources that was placed nearby as a reference.

3. Characterization of two 1" x 1"CLYC scintillators

Two crystals were used in this work: a CLYC enriched at 95% of ^{6}Li (CLYC-6) and a CLYC with an enrichment of ^{7}Li larger than 99% (CLYC-7). Both crystals were produced by RMD and have a cylindrical shape, a diameter of 1" and a thickness of 1". By visual inspection the crystals present some small internal structures. However, they do not seem to affect the energy resolution. These two samples of CLYC scintillators were coupled to a HAMAMATSU R6231-100mod PMTs and to standard voltage dividers (HAMAMATSU E1198-26 and HAMAMATSU E1198-27 for CLYC-6 and CLYC-7, respectively).

3.1. General Properties

The CLYC scintillators exhibit good energy resolution. We measured 4.8% and 4.5% at 662 keV for CLYC-6 and CLYC-7, respectively. The CLYC scintillators can identify and measure gamma-rays

Journal of Physics: Conference Series **620** (2015) 012003 doi:10.1088/1742-6596/620/1/012003

and neutrons via pulse shape discrimination. The CLYC internal radiation was measured to be at least two order of magnitude smaller than that of an equivalent $LaBr_3$:Ce detector [8].

3.2. Thermal and fast neutron identification

The thermal neutrons were measured with both detectors, using an AmBe source. The spectra of Fig. 5 and Fig 6 were acquired in the same conditions and are normalized on the ^{137}Cs peak. The 3.2 MeV peak, induced by thermal neutrons, is practically absent in the CLYC-7 spectrum (see Fig. 6), but it is visible in the CLYC-6 one (see Fig. 5). The thermal neutron detection efficiency for the CLYC-7 is 0.3% with respect to the CLYC-6 sample of identical size.

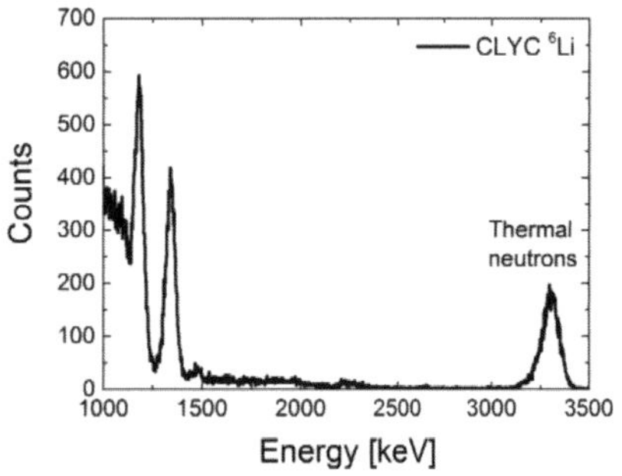

Figure 5. The energy spectra measured with CLYC-6 and a AmBe(Ni) source in a paraffin box. A ^{60}Co and a ^{137}Cs were also present.

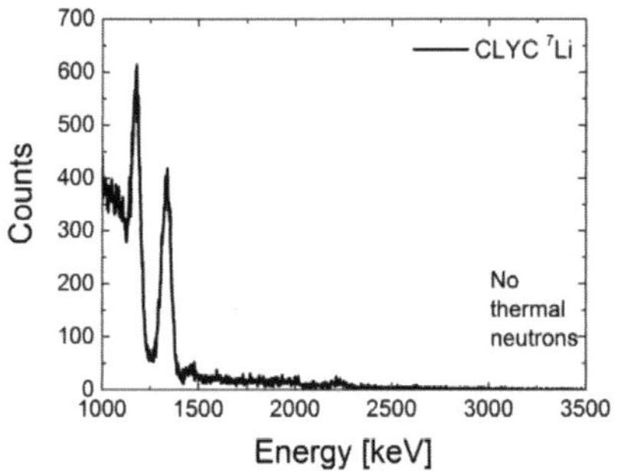

Figure 6. The energy spectra of CLYC-7 acquired in the same condition of the spectra of Fig. 5.

Both crystals are capable to detect, identify and measure the kinetic energy for fast neutrons. The ability to separate between neutron and gamma-rays does not depend on the neutron energy while the measurement of neutron kinetic energy using the CLYC scintillators can be performed up to approximately 8 MeV. In the case of neutrons of higher kinetic energy three body reactions have a larger cross section of the reaction on ^{35}Cl. A continuum energy spectrum was measured caused by a superimposition of different reaction mechanisms [8].

3.3. PSD with different PMTs

The CLYC light emission has a fast and slow component related. The fast component, manly in the UV-blue, is significant only when gamma rays are measured and it may be used for effective pulse shape discrimination. Timing or spectroscopic PMTs provide different responses to a fast signal because of their intrinsic rise time; in addition, the entrance window (quartz or borosilicate glass) filters in a different way the CLYC fast emissions.

The CLYC-6 crystal was coupled to the different PMTs and irradiated using gamma-rays sources and the AmBe source. The used PMTs are all from HAMAMATSU: i) H6533 (timing PMT with borosilicate glass window); ii) R2059 (timing PMT with quartz window); iii) R6233-100sel (spectroscopic PMT with borosilicate glass window) and iv) R6231-100mod spectroscopic PMT with quartz window). The fast component is more visible for PMTs with fast timing response (R2059 and H6533). A figure of merit (FOM), defined as the difference between the centroid of the neutron and gamma peak divide by the sum of the FWHM of the two peak, by projecting a PSD matrix, like the one of Fig. 7, along the y axes. The FOM describes the capability to discriminates between gamma rays and neutrons and it ranges from 2.8 to 3.4 was found. It is important to point out that the FOM of a ^{3}He tube is ~ 2. In optimized conditions using custom electronics and digital filters FOM of 3.8 was found for a HAMAMATSU R6231-100 PMT, as shown in Fig. 7 [8].

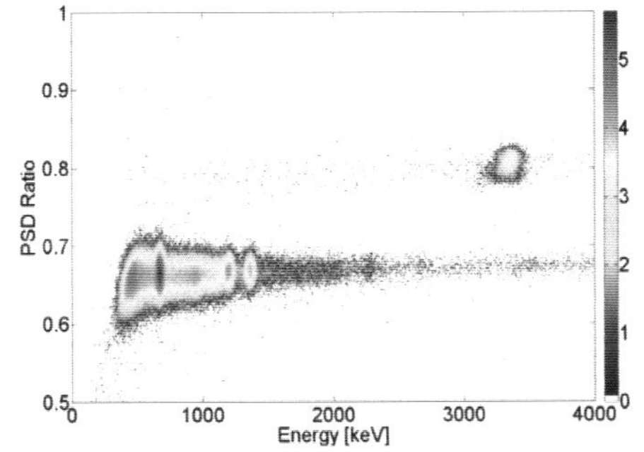

Figure 7. The PSD matrix of CLYC-6 coupled with an HAMAMATSU R6231-100mod PMT. The logarithm of the counts of the matrix is plotted on the z axis. The measured FOM is 3.8.

4. Conclusions

In this work, GYGAG:Ce (0.3" x 2"), SrI_2:Eu (2" x 2") and $CeBr_3$ (3" x 2") samples were characterized by studying the pulse shape, the energy resolution and the crystal responses to a collimated source. The SrI_2:Eu has the best energy resolution but it presents a strong self-absorption. The $CeBr_3$ has no internal activity and a fast signal, similar to $LaBr_3$:Ce. The GYGAG:Ce seems to be a very good detector, with high efficiency, good energy resolution and the possibility to be produced in every dimension and shape. Moreover, two samples (enriched with ^{6}Li and ^{7}Li) of CLYC scintillators were tested with gamma rays, thermal and fast neutrons. The performances of CLYC scintillators with different PMT were measured.

References

[1] Cherepy N., *et al.*, 2013, *IEEE Tran. on Nucl. Science*, 60, 955-958.
[2] Quarati F. G. A., Dorenbos P., van der Biezen J., Owens A., Selle M., Parthier L and Schotanusf P., 2013, *Nucl. Instr. and Meth. A:*729, 596-604
[3] Cherepy N., *et al.*, 2013, *IEEE Tran. on Nucl. Science*, 60, 2330-35.
[4] Glodo J., Hawrami R. and Shah K.S., 2013 *J. Crys. Growth*, 379, 73–78
[5] D'Olympia N., Chowdhury P., Lister C. J, Glodo J., Hawrami R., Shah K. and Shirwadkar U., 2013, *Nucl. Instr. and Meth, A* 714, 121-127.
[6] Riboldi S., *et al.*, 2011, *Nucl. Science Symp. Conf. Record* (NSS/MIC), 776-778.
[7] Giaz A., Hull G., Fossati V., Cherepy N., Camera F., Blasi N, Brambilla S., Coelli S., Million B. and Riboldi S., summitted to *IEEE Trans. on Nucl. Science.*
[8] Giaz A. *et al.*, in preparation to be submitted to *Nucl. Instr. Meth. A.*

Journal of Physics: Conference Series **620** (2015) 012004 doi:10.1088/1742-6596/620/1/012004

Position sensitivity in 3"x3" Spectroscopic LaBr3:Ce Crystals

N.Blasi, A.Giaz, C.Boiano, S.Brambilla, F.Camera, B.Million, S.Riboldi

INFN- Sezione di Milano, Italy

Physics Department, University of Milano, Italy

Abstract. The position sensitivity of a thick, cylindrical and continuous 3" x 3" (7.62 cm x 7.62 cm) $LaBr_3$:Ce crystal with diffusive surfaces was investigated. Nuclear physics basic research uses thick $LaBr_3$:Ce crystals (> 3cm) to measure medium or high energy gamma rays (0.5 MeV < E_γ< 20 MeV). In the first measurement the PMT photocathode entrance window was covered by black absorber except for a small window 1 cm x 1cm wide. A complete scan of the detector over a 0.5 cm step grid was performed. The data show that even in a 3" thick $LaBr_3$:Ce crystal with diffusive surfaces the position of the full energy peak centroid depends on the source position. The position of the full energy peak centroids are sufficient to identify the collimated gamma source position. The crystal was then coupled to four Position Sensitive Photomultipliers (PSPMT). We acquired the signals from the 256 segments of the four PSPMTs grouping them into 16 elements. An event by event analysis shows a positon resolution of the order of 2 cm.

1. Introduction

The position sensitivity of a thick, cylindrical and continuous 3" x 3" (7.62 cm x 7.62 cm) $LaBr_3$:Ce crystal with diffusive surfaces was investigated. The $LaBr_3$:Ce material is an inorganic scintillator which presents excellent scintillation properties. Its position sensitivity properties were studied and applied by many groups in several fields, mostly medical. Its high light yield allows a sub-millimeter position resolution in gamma-camera for small animal SPECT. Usually, those gamma cameras are used to detect gamma rays of known energy, mostly the 140 keV from the ^{99m}Tc source, or the 511 keV coming from positron annihilation, in order to locate the position and dimension of the emitter with a position resolution as good as possible. Following these requirements, crystals used in this field are mostly continuous, with a maximum thickness of the order of 1 cm, and they have absorbing surfaces, so that only the scintillation light coming directly from the interaction point reaches the photocathode. In nuclear physics basic research, $LaBr_3$:Ce crystals are used to study the decay properties of nuclei populated via nuclear reactions, where an accelerated particle hits a target nucleus. Therefore, the position of the emitter is well known, while the energies of the gamma transitions are unknown a priori, and they could range up to several MeV.

With the availability of exotic beams, nuclei far from stability can be studied via inverse reactions where the nucleus under study is accelerated, excited by the interaction with a target nucleus and de-excited by radiation emission while still moving. These beams can be accelerated to velocities up to v/c = 0.7-0.8. In the lab system, due to the Doppler effect, the energy of a gamma ray emitted by a moving source depends on the angle of emission. Therefore, the energy detected will be degraded

Journal of Physics: Conference Series **620** (2015) 012004 doi:10.1088/1742-6596/620/1/012004

depending on the solid angle covered by the detector. For example, in a 3"x3" $LaBr_3$:Ce detector placed at 20 cm distance from the target, a 1 MeV gamma ray will be detected with 25 kev resolution if the emitter is at rest, 70 keV resolution if it moves with v/c=0.3 or 230 keV for v/c =0.7. The problem of Doppler broadening is so important in nuclear physics that arrays have been developed [1,2], made of segmented germanium crystal, able to track the detected gamma rays, to identify the interaction point, and correct the gamma energy for the angle of emission [3]. The advantage of these arrays is an excellent energy resolution, but the system is very complicated, hardly transportable from one lab to the other, the setup is complex, and also the data analysis is quite laborious. Last but not least, it is very expensive.

On the other hand, it is not always necessary to have such an excellent energy resolution, for example when studying giant resonances at higher excitation energies, 3 or 4% energy resolution would be sufficient [4]. Our aim is to investigate for a position sensitive detector with a good efficiency, for energies up to several MeV, good energy resolution, simple to handle and to transport, and as cheap as possible. $LaBr_3$:Ce crystal are ideal candidates, since they are available in large volumes and have good energy resolution when provided with diffusive surfaces, so that all the scintillation light is collected. Since we use imaging to recover energy resolution, considering the intrinsic resolution of the $LaBr_3$:Ce, a spatial resolution of the order of 1 cm would be sufficient.

2. Statistical measurement

There are two big differences between a nuclear physics spectroscopic detector and the ones usually used for imaging: the first is the thickness. We do not expect to find position sensitivity for one or two hundred keV gamma rays, since they all interact in the first millimeters of the scintillator [5], but for medium or high energies gamma rays position sensitivity might be retained. The second difference is the presence of diffusive surfaces. A 1 MeV gamma ray produces about 60,000 scintillation photons, the largest part of which will undergo several reflections on the surfaces covering a mean path of the order of 1 meter before hitting the photocathode. Therefore the light distribution on the detection surface might be heavily affected.

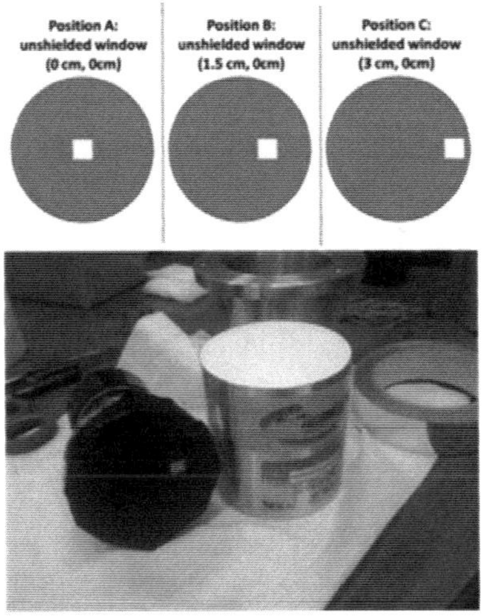

Figure 1. Lower panel: the shielded PMT together with the 3"x3" $LaBr_3$:Ce crystal is shown. Upper panel: the position of the three windows in the PMT shielding are indicated: from the left: position A (0cm,0cm); position B (1,5cm,0cm); position C (3cm;0cm)

To verify if position sensitivity is still present, we performed a simple test, taking a Hamamatsu R6233-100SEL spectroscopic PMT 3"X3" which, coupled to a Hamamatsu E1198-26 voltage divider usually gives 20 keV resolution at 662 keV. We shielded the tube with black tape, leaving a square window with 1 cm side (see Fig. 1). We then collimated an intense ^{137}Cs source into a beam spot with a diameter of 1 mm and performed a set of measurements moving the source on a half a centimeter grid. The measurements were repeated for three window positions: in the center, at 1.5 and 3 cm from the center. As a result, by plotting the centroids of the 662 keV peak as a function of the source position, we find a correlation between the window position and the source position [6] (see Fig. 2).

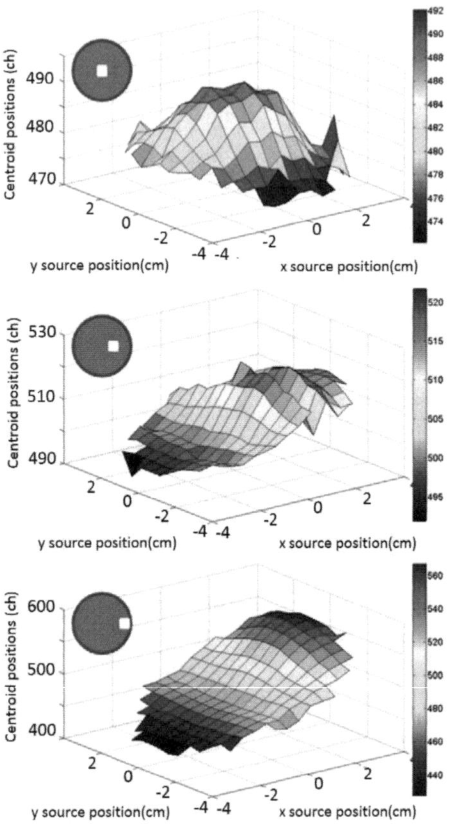

Figure 2. Centroids of the 662 keV peak are plotted as a function of the source position for the three window positions, as indicated in the insets.

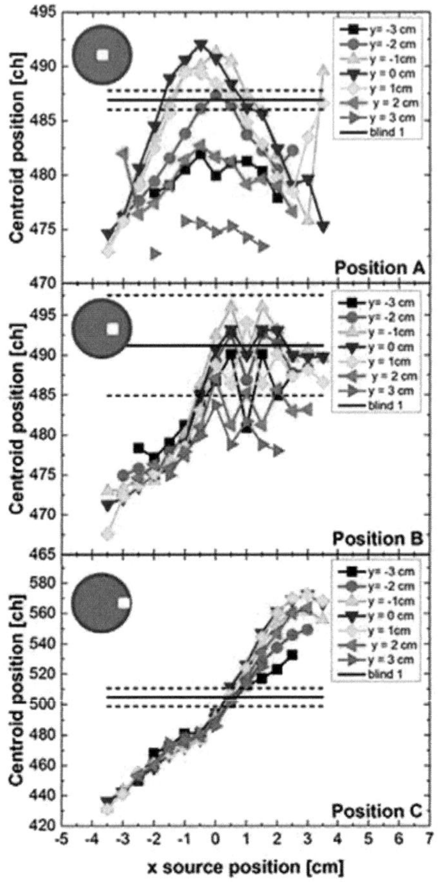

Figure 3. The positions of the full energy peak centroid obtained in the measurements with the source in the "unknown" position (black lines) are compared with the tabulated centroids as a function of the source position along the x axis for the three window positions, as indicated in the insets. The error bars of the "blind" measurements are represented by the dashed black lines.

We can conclude that some position sensitivity is still present. In order to check if this is enough to extrapolate the source position from the data, we tabulated the centroid values corresponding to the grid source positions for each window. An additional table relative to a fourth window position was deduced by applying a 90° rotation to the window at 3 cm from the center. We then placed the source in an "unknown" position and performed four measurements with the four window positions in the shield.

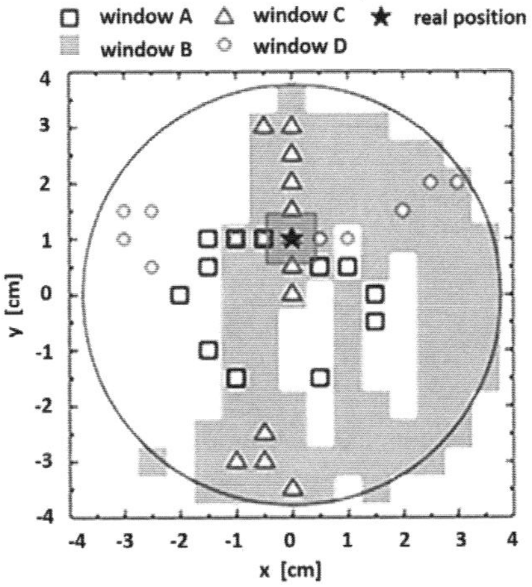

Figure 4. The coordinates corresponding to the measured centroids overlapping with the tabulated ones are labelled with different symbols for each window position. The circle represents the detector front face. The real source position is indicated by the star.

The measured centroids were compared to the tabulated ones and the source coordinates corresponding to the centroids overlapping with the data were selected (see Fig. 3). They were plotted on a plane, representing the crystal front face, as shown in Fig. 4. The overlap of the selected coordinates corresponds to the source position and we find a good agreement within $1cm^2$, which is the grid step dimension. These results are very promising.

3. Event by event measurement

For practical application, we need an event by event analysis. This requires the use of segmented position sensitive photo tubes. We used the Hamamatsu H8500, which is the largest available, it has dimensions 2"x2" and 64 segments. In order to cover the whole crystal surface, we used four of such tubes. They were not equal: two of them had 10 dynodes and a bialkal photocathode, while the other two had 8 dynods and SBA photocathodes. The performances of the two types of PMT's were different: the energy resolution of each single PMT coupled to the crystal was about 5 % for the first type and about 4% for the other. When used together, the energy resolution of the four PMT's was 4% at 662 keV. The complete cover of the detector surface implies the use of 120 segments. In order to have a simpler setup, we short circuited segments in groups of 16, so that we ended up with 12 macro segments, as shown in fig. 5.

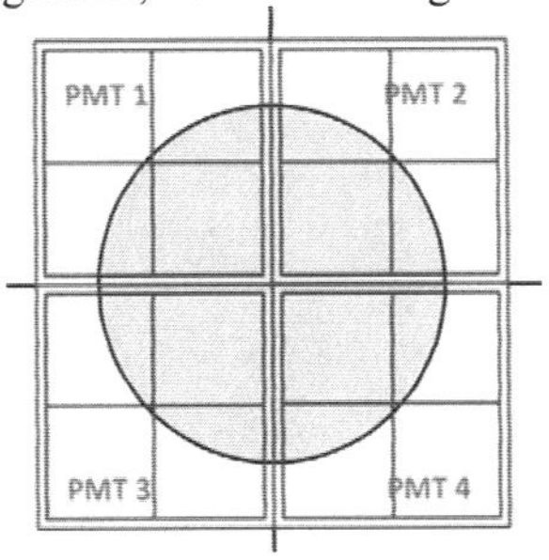

Figure 5. The 12 macro segments covering the crystal surface, obtained by short circuiting groups of 16 elements of the PMT's, are shown. The four corner macro segments were not considered since the covered surface was negligible.

Data were taken with the collimated ^{137}Cs source in several positions, which are indicated by open squares in fig. 6. The coordinates of each event were obtained calculating the center of gravity of the light distribution. Gating on the 662 keV transition, the position profiles are Gaussian distributions with a FWHM of about 2.3 cm. If we plot the centroids of the distributions as a function of the source position, we see a deviation from linearity as found in all imaging studies (Fig.7). By correcting for

Journal of Physics: Conference Series **620** (2015) 012004 doi:10.1088/1742-6596/620/1/012004

this non linearity performing a fit with a 3rd degree polynomial (solid line in Fig. 7), the source position is very well reproduced, as can be seen in Fig. 6.

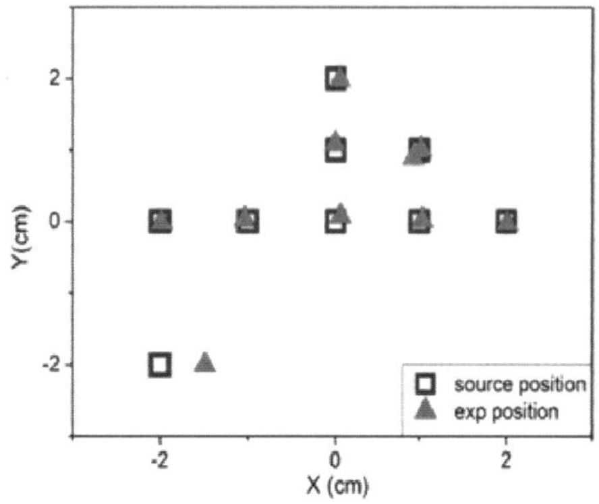

Figure 6. The calculated source positions (triangles) are compared to true position of the source (open squares).

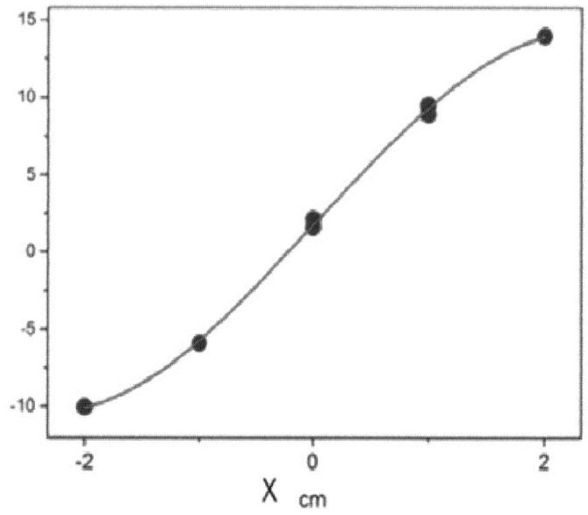

Figure 7. The graph shows the measured linearity along the x-axis. The line is a polynomial fit.

4. Conclusions

The position sensitivity of a spectroscopic 3"x3" LaBr3:Ce detector was investigated through measurements using an intense ^{137}Cs collimated beam of 1 mm diameter. A simple test using a Hamamatsu R6233-100SEL spectroscopic PMT shielded so that only a 1 cm^2 is left shows a correlation between source and window positions. The event by event analysis, performed with Position Sensitive PMT's, demonstrates a position sensitivity with a resolution of the order of 2 cm for 662 keV gamma rays. Such resolution would be sufficient to correct for Doppler broadening in nuclear physics experiments with exotic beams.

References

[1] M. Deleplanque et al., Nucl. Instr. Meth. Phys. Res. A, Vol. 430, 292, 1999.
[2] S. Akkoyun et al., Nucl. Instr. Meth. Phys. Res. A, Vol. 668, 26, 2012.
[3] F.C.L. Crespi et al., Nucl. Instr. Meth. Phys. Res. A, Vol. 705, 47, 2013.
[4] O. Wieland et al., Phys. Rev. Lett., Vol. 102, 092502, 2009.
[5] J.W. Scrimger and R.G. Baker, Phys. Med. Bio., Vol. 12, 101, 1967
[6] A. Giaz et al., Nucl. Instr. Meth. Phys. Res. A, Vol. 772, 103, 2015.

Development of NANA: A Fast-Scintillator, Coincidence Gamma-ray Array for Radioactive Source Characterisation and Absolute Activity Measurements at the UK National Physical Laboratory

P.H.Regan[1,2], R.Shearman[1,2], S.M.Judge[1,2], G.Lorusso[1,2], P.Main[1,2], S.Bell[1], S.M.Collins[1], P.Ivanov[1], S.M.Jerome[1], J.D.Keightley[1], C.Larijani[1], G.Lotay[1,2], A.K.Pearce[1]

National Physical Laboratory, Teddington, Middlesex, TW11 0LW, UK
Department of Physics, University of Surrey. Guildford, GU2 7XH, UK

E-mail: p.regan@surrey.ac.uk ; paddy.regan@npl.co.uk

Abstract. A multi-detector modular coincidence gamma-ray spectrometer is being designed and constructed for use at the UK's National Physical Laboratory (NPL) for use in direct measurement and metrological standardisation of nuclear decay activities. In its first generation, the NPL **Na**tional **N**uclear **A**rray (NANA) will consist of twelve individual halide scintillation detectors placed in a high-efficiency geometry around a well-defined central point source position. This brief conference paper provides details of the measured detector module and coincidence energy and timing responses for the LaBr3(Ce) detectors which will be used in the NANA array. Preliminary GEANT4 simulations of the array's full energy peak efficiency and expected gamma-ray coincidence response are also presented.

1. The National Nuclear Array - NANA

The UK National Physical Laboratory, NPL is the UK's national standard's laboratory and underpins the UK's national Measurement system. NPL has responsibility for the traceability of the unit of radioactive decay, namely the Becquerel (Bq) and for standardizations of the decay characteristics for a wide range of radioactive species. To this end, a novel, multi detector gamma-ray spectrometer array is currently under construction to be based at NPL. It is anticipated at the *National Nuclear Array* or NANA, will allow traceable characterizations of radionuclide species activities with complex nuclear decay schemes using gamma-ray energy coincidences. This will allow standardization of defined radioactive species and also act as a traceable primary radiation standard. The proposed array could also be used for nuclear material assay and the identification of which nuclear species are present in a given sample for nuclear forensics studies through the identification of discrete-line characteristic gamma-ray signatures and their associated specific activities. Examples of potential nuclear material which could be characterized using the NANA spectrometer include radiopharmaceutical tracers such as ^{177}Lu [1] and decay products from nuclear waste materials, in particular, complex, energy spectra arising from sources of mixed, long-lived nuclear fission fragments [2,3]. The array could also be used for contributions in basic, fundamental science such as investigations of nuclear structure and nuclear decay data measurement by evaluations of the nuclear energy level schemes for specific radionuclide decays, such as the members of the naturally occurring radioactive decay chains headed by uranium and thorium isotopes [4]. This short conference paper will outline some of the design parameters and expected spectroscopic performance and response of NANA and its detector components.

2. $LaBr_3$(Ce) Detector Performance and Coincidence Measurement.

The idealized gamma-ray spectrometer array combines the best aspects of both nuclear spectrometry and decay calorimetry. To this end, a gamma-ray spectrometer for use in nuclear decay measurements has the best combination of [5]:

- discrete-line, energy resolution, in order to be able to select out cleanly individual gamma-ray transitions from a large background of other transitions;
- Full-energy peak efficiency over a wide range of likely gamma-ray energies. This both reduces counting times and enables the identification of signature decay lines from weakly concentrated and/or long half-life nuclides.
- To have good timing characteristics, ideally with a sub-nanosecond timing response. This allows clean correlations between two discrete gamma-rays in mutual coincidence. It also allows the measurement of transitions rates of nuclear excited states in the nanosecond regime.

No individual gamma-ray detection material is idealized in each of these three areas, with HPGe detectors having superior energy resolution and heavier scintillation materials such as BGO giving the largest full-energy peak efficiency but at the price of poor energy resolution. The proposed NANA array will possibly house 12 individual $LaBr_3$(Ce) detectors of the geometry used by the FATIMA collaboration for the DESPEC nuclear fast-timing project at NUSTAR [6,7]. The $LaBr_3$(Ce) crystal dimensions in each case are 2” in length and 1.5” diameter cylinders, housed in an aluminium holding can and attached to a Hamamatsu R9779 fast-timing photomultiplier tube (PMT).

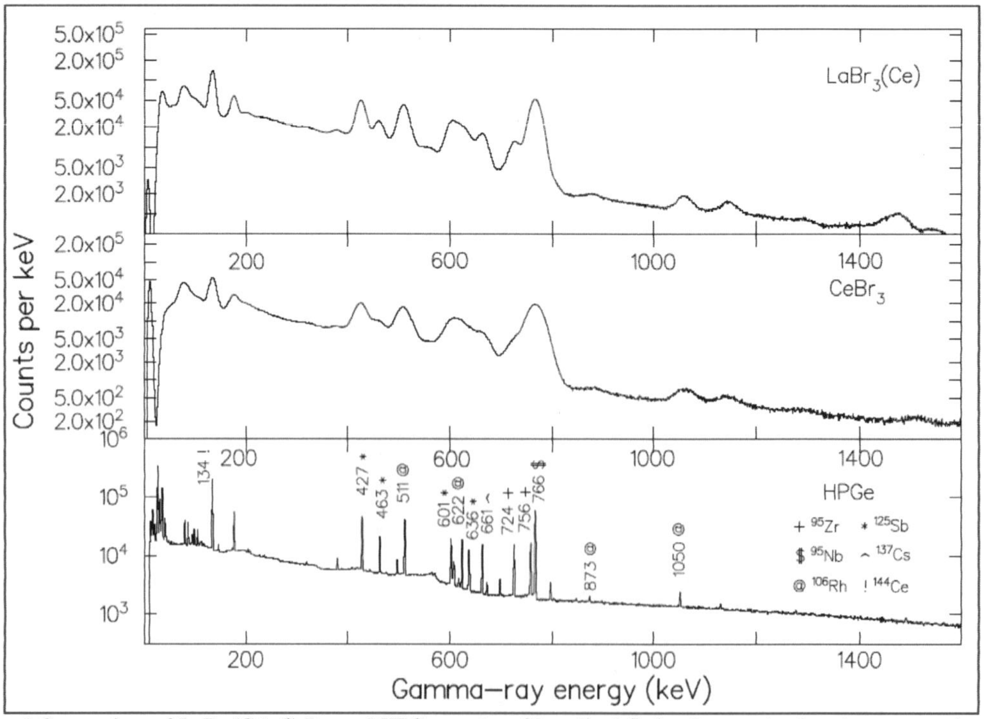

Figure 1 Comparison of $LaBr_3$(Ce), $CeBr_3$ and HPGe spectra of long-lived fission fragments of protons on uranium. The main lines identified in the HPGe spectra arise from the decays of ^{95}Zr, ^{95}Y, ^{106}Rh, ^{125}Sb, ^{137}Cs and ^{144}Ce decays. See reference [2] for source preparation details.

In the last five years, a range of examples of nuclear spectroscopic measurements have been presented in the literature using halide scintillation materials, specifically Cerium-doped Lanthanum tri-bromide, $LaBr_3$(Ce), for spectroscopic studies [8-21]. Figure 1 shows a comparison of the spectral responses for HPGe detectors and also two halide scintillation detectors, namely $LaBr_3$(Ce) and $CeBr_3$ for a long-lived source of fission products produced in the proton-induced fission of ^{238}U. We conclude that $LaBr_3$(Ce) combines acceptable energy resolutions over a wide range of gamma-ray energies, reasonable linearity of response and acceptable full-energy peak efficiency for a national standardization instrument, such as NANA.

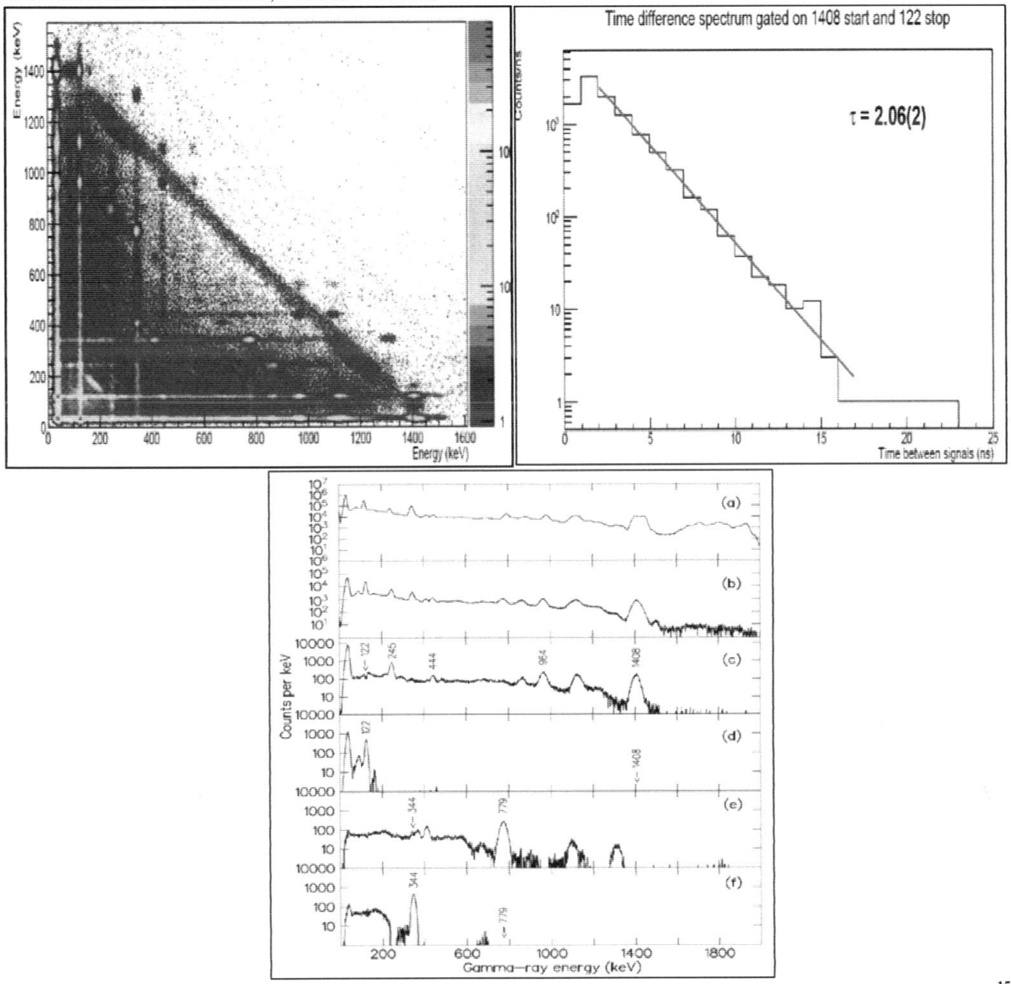

Figure 2 (Upper Left) 2-D gamma-ray energy coincidence matrix between two $LaBr_3$ detectors viewing a ^{152}Eu radioactive point source with a 25 ns time overlap timing requirement. Note the clearly separated correlated events between the 344 keV and 779 keV gamma-ray transitions from the decay into ^{152}Gd and those transitions in coincidence with the 122 keV $2^+ \rightarrow 0^+$ transition in the decay branch to ^{152}Sm. (Upper right) Time difference spectrum using the CAEN V1751C 1 GHz timing signal between the detection of the 1408 keV and 122 keV coincident transitions across the $I^{\pi}=2^+$ state in ^{152}Sm populated from the ^{152}Eu radioactive source. The fitted line is a single component exponential decay for the mean-lifetime of the intermediate nuclear state. The fitted value gives a mean-lifetime for the decay of the I^{π}=2- state in ^{152}Sm of 2.06(2) ns, which compares with the ENSDF value for this lifetime is 2.025 (16) ns [26]. (Lower Panel) Gamma-ray spectra from the same source showing (a) $LaBr_3$ singles spectra; (b) total projection of the gamma-ray coincidence matrix with a requirement that the two gamma rays were measured within 25 ns of each other; (c) gate on the 122 keV gamma ray full-energy peak (from decays into ^{152}Sm); (d) gate on the 1408 keV transition (^{152}Sm); (e) gate on 344 keV $2^+ \rightarrow 0^+$ in ^{152}Gd; and (f) gate on 779 keV transition (^{152}Gd). A fraction of the total projection spectrum has been subtracted channel by channel from the gated spectra.

We have performed a range of additional evaluations on the performance of the $LaBr_3(Ce)$ detector modules used in the FATIMA set-up. Figure 2 shows the gamma-ray energy coincidence matrix for a ^{152}Eu source placed between two cylindrical FATIMA $LaBr_3$ detectors of length 1.5" and diameter 2". The time-stamped energy signals for both detectors were measured using the CAEN V1751C, 1 GHz, 8 channel module to test its time response and performance. Note the improvement in peak-to-background ratio for examples where a clean gamma-ray coincidence correlation requirement is applied between the two $LaBr_3$ detectors. Simple coincidence requirements on (a) the 122 keV gamma ray to select the ^{152}Sm decay cascade and (b) the 344 keV transition to select the competing ^{152}Gd cascade have been applied in off-line analysis of the event by event data. The increase in the sensitivity of transitions in cascade coincidence with the 122 keV decay in ^{152}Sm (e.g., 244, 444, 964 and 1408 keV) is clear and demonstrates the potential effectiveness of coincidence spectrometry of source materials for characterization and standardizations using NANA. The time difference between the 122 keV and 1408 keV transitions across the first excited state in ^{152}Sm using the simple 1 GHz time stamping of the signals using the CAEN digitizers gives a mean-lifetime for this state of 2.06(2) ns, which is consistent with the evaluated value.

3. GEANT4 Simulations of Possible NANA Geometry.

The design of NANA in its first iteration is based on a range of GEANT4 simulations [22] of array response for standard sources for up to 12 $LaBr_3(Ce)$ detector modules of the dimensions used in the FATIMA gamma-ray spectrometer [6] which is part of the DESPEC project within the NUSTAR collaboration [7,18]. Figure 3 shows a GEANT4 geometry the proposed design for the NANA array, which covers approximately 22% of the total solid angle for a source placed in the centre of the array. Using GEANT4 simulations, which assume a realistic energy resolution of 3.5% at 662 keV, the expected full-energy peak efficiency for the proposed array geometry corresponds to approximately 3% at a gamma-ray energy of 1332 keV.

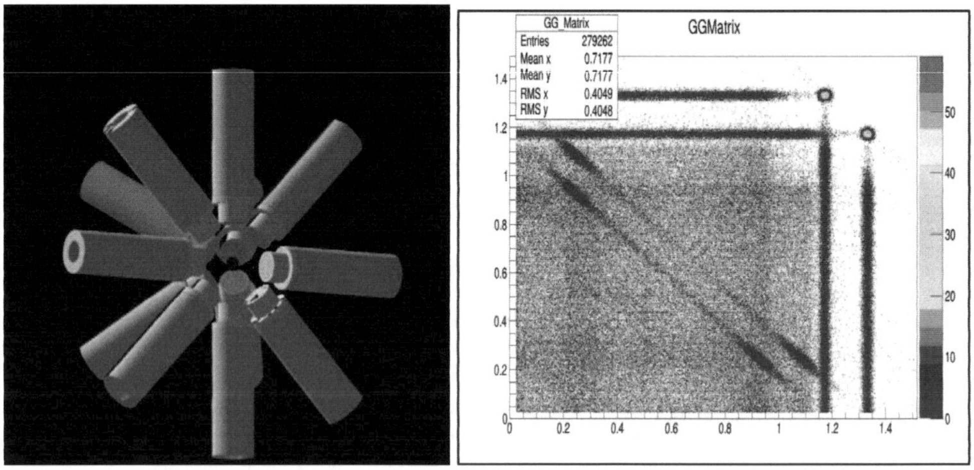

Figure 3: (Left): GEANT4 geometry for NANA consisting of 12 $LaBr_3(Ce)$ FATIMA module detectors, with 8 modules in a central ring at a distance of 90.5 mm from the central target position and four at the back plane at a distance of 50 mm from the target position. (Right) Results of a GEANT4 simulation of 10^7 individual ^{60}Co decays for a point source at the centre of the using the NANA design proposed. The array shows the gamma-ray energy coincidence response of the array. Note that the diagonal lines correspond to summed Compton scattered events between two separate detectors. The axes are labelled in gamma-ray energy in units of MeV.

4. Summary and Future Plans.

The design of a gamma-ray coincidence spectrometer for radioactive source standardization and characterization, NANA, have been undertaken. Characterization of the 1.5" diameter x 2" long

Journal of Physics: Conference Series **620** (2015) 012005 doi:10.1088/1742-6596/620/1/012005

cylindrical $LaBr_3$(Ce) detectors from the FATIMA collaboration has been carried out and their response compared with both HPGe and 1.5"x1.5'' $CeBr_3$ detector modules. A time-stamped, digital, data acquisition system has been evaluated and used in gamma-ray coincidence mode to evaluate the effect of discrete-line gamma-ray energy coincidences by differentiating between different decay paths from a ^{152}Eu standard point source between two $LaBr_3$ detectors. GEANT4 simulations have been carried out for the initial design criteria of the NANA spectrometer have been carried out and design consisting of 12 individual FATIMA LaBr3(Ce) modules, resulting in an expected full-energy peak gamma-ray detection efficiency of ~3% at 1.3 MeV. Future investigations and modelling of potential updates for NANA, which might include the inclusion of additional $CeBr_3$ [23,24] and/or HPGe detectors for gamma-ray calorimetry, gamma-ray multiplicity measurements, sub-nanosecond nuclear timing measurements and gamma-ray coincidence gating selection will be investigated in future studies. The response of NANA for gamma-ray energies of up to 20 MeV will also be evaluated [25].

Acknowledgements
This work is funded through the UK National Measurement Office. RS acknowledges receipt of a PhD studentship from the NDA/NNL. PHR acknowledges additional financial support from the UK Science and Technology Facilities Council. This work is partially supported by the METRO ERM and IND-57 METRONORM projects funded by the EMRP.

References

[1] F.G.Kondev et al., Applied Radiation and Isotopes **70**, 1867 (2012); E.A.Rezende et al., Applied Radiation and isotopes **70**, 2081 (2012)
[2] C.Larijani et al., Radiation Physics and Chemistry (2015) in press, http://dx.doi.org/j.radphyschem.2015.01.021
[3] S.Bender, B.Heidrich and K.Unlu, Nuclear Instruments and Methods in Physics Research A, in press, (2014), http://dx.doi.org/10.1016/jnima.2014.12.025
[4] S.M.Collins et al., Applied Radiation and Isotopes, **99**, 46 (2015)
[5] J.Eberth and J.Simpson, Progress in Particle and Nuclear Physics **60**, 283 (2008); C.W.Beausang and J.Simpson Journal of Physics G: Nuclear and Particle Physics **22**, 527 (1996)
[6] O.Roberts et al., Nuclear Instruments and Methods in Physics Research A **748,** 91 (2014)
[7] P.H.Regan, Applied Radiation and Isotopes, **70**, 1125 (2012)
[8] J.M-Regis et al., Nuclear Instruments and Methods in Physics Research A **684**, 36 (2012)
[9] J.M-Regis et al., Nuclear Instruments and Methods in Physics Research A **622**, 83 (2010)
[10] T.Alharbi et al., Applied Radiation and Isotopes **70**, 1337 (2012)
[11] S.Zhu et al., Nuclear Instruments and Methods in Physics Research A**652**, 231 (2011)
[12] P.J.R.Mason et al., Physical Review C**85**, 064303 (2012)
[13] T.Alharbi et al., Physical Review C**87**, 014323 (2013)
[14] P.J.R.Mason et al., Physical Review C **88**, 044301 (2013)
[15] J.M.Regis et al., Nuclear Instruments and Methods in Physics Research A**763**, 210 (2014)
[16] J.M.Regis et al., Physical Review C **90**, 067301 (2014)
[17] T.Alharbi et al., Physical Review C **91**, 027302 (2015)
[18] P.H.Regan et al., European Physical Journal Web of conferences **63**, 01008 (2013)
[19] D.L.Smith et al., Phys. Rev. C**77**, 014309 (2008)
[20] N.Marginean et al., European Physical Journal A **46**, 329 (2010)
[21] S.Kisyov et al., Phys. Rev. C**84**, 014324 (2011)
[22] S.Agostineli et al., Nuclear Instruments and Methods in Physics Research A **506**, 250 (2003)
[23] N.D'Olympia et al., Nuclear Instruments and Methods in Physics Research A **728**, 31 (2013)
[24] L. Fraile et al., Nuclear Instruments and Methods in Physics Research A **701**, 235 (2013)
[25] M.Ciemala et al., Nuclear Instruments and Methods in Physics Research A **608**, 76 (2009)
[26] M.J.Martin Nuclear Data Sheets **114**, 1497 (2013)

The PARIS cluster coupled to the BaFPro electronic module: data analysis from the NRF experiment at the γELBE facility

B Wasilewska[1], P Bednarczyk[1], C Boiano[2], S Brambilla[2], F Camera[2,3], M Ciemała[1,4], O Dorvaux[5], A Giaz[2], M Jastrząb[1], S Kihel[5], M Kmiecik[1], A Maj[1], I Matea[6], R Massarczyk[7], I Mazumdar[8], A Mentana[2], P Napiorkowski[9], B Sowicki[1], R Schwengner[7], S Riboldi[3], M Ziębliński[1] and the PARIS collaboration

[1] IFJ PAN Cracow, Poland, [2] INFN Milano, Italy, [3] University of Milano, Italy, [4] GANIL Caen, France, [5] IPHC Strasbourg, France, [6] IPN Orsay, France, [7] HZDR Dresden, Germany, [8] TIFR Mumbai, India, [9] SLCJ Warsaw, Poland

E-mail: barbara.wasilewska@ifj.edu.pl

Abstract. The first cluster of the constructed PARIS calorimeter was assembled and tested at the γELBE facility at HZDR, Dresden, Germany. The experiment was aimed at the evaluation of the performance of each detector separately as well as the whole PARIS cluster with discrete γ-ray energies seen by the PARIS ranging up to 8.9 MeV. As the detectors use *phoswich* configuration, with 2'' x 2'' x 2'' $LaBr_3$(Ce) crystal coupled to 2'' x 2'' x 6'' NaI(Tl) one, great care must be taken during the data analysis process to obtain the best possible values for energy resolution. Two algorithms for data transformation from matrices created with slow vs fast pulse shaping to energy spectra were tested from which one was chosen for further analysis. An algorithm for adding back energies of γ-rays scattered inside the cluster was prepared, as well. Energy resolution for γ-rays in 2–8 MeV range was estimated and is presented in this paper.

1. Introduction

The **P**hoton **A**rray for studies with **R**adioactive **I**on and **S**table beams (PARIS [1]) is a new 4π calorimeter whose properties combine high detection efficiency and energy resolution for γ-rays in energy range from 500 keV to 50 MeV. It will be made of *phoswich* detectors, consisting of 2'' x 2'' x 2'' crystals of a new generation scintillator of $LaBr_3$(Ce) backed by 2'' x 2'' x 6'' crystal of standard material of NaI(Tl) with a common PMT, arranged in clusters of nine (Fig. 1). This configuration ensures high efficiency, yet, as light impulses emitted from those materials differ significantly, resulting in distinguishable from each other signals shapes, Pulse Shape Analysis (PSA) methods must be engaged. Additionally, an event of the 511 keV γ-rays escape from a crystal is common in high γ-ray energy range, which calls for an extra off-line procedure to suppress that effect.

This array will be used in a number of nuclear physics cases spanning from time measurements of weak discrete transitions to measurements of properties of exotic collective modes such as Giant Quadrupole Resonances [2], [3] and Pygmy Dipole States [4], [5]. What all those

Journal of Physics: Conference Series **620** (2015) 012006 doi:10.1088/1742-6596/620/1/012006

experiments will have in common are complexity and high exclusiveness of the events of interest. Thus, it is crucial to fully characterise the PARIS performance and prepare algorithms and methods, which will help obtain the best possible results, beforehand.

2. Experimental setup

During the Nuclear Resonance Fluorescence (NRF) experiment at the γELBE facility [6] at HZDR, Dresden, Germany, a 16 MeV electron beam was converted into Bremsstrahlung photons, which subsequently irradiated the target of 4 g of ^{11}B. In this way mainly the states at 2125, 4445, 5020, 7285 and 8920 keV were excited and the respective transitions to the ground-state were measured. The PARIS cluster was placed at the backward angle of 125° and the detectors were shielded from the beam with the use of lead blocks. At the front, absorbers made of copper and lead were mounted. The used PMT coupled with each PARIS *phoswich* was Hamamatsu R7723-100.

Data were taken using an analogue BaFPro [7] electronic module which has two branches of pulse shaping: fast and slow, corresponding to amplitude and charge of the impulse respectively. Both signals were collected and obtained information was used to separate three kinds of events (with full photon energy deposition in $LaBr_3$(Ce), in NaI(Tl) and partial deposition in both materials). As an example a matrix containing data obtained with fast and slow pulse shaping of signals with marked its crucial properties is presented in Fig. 2. There were two time information sets, derived from both devices independently. Standard radioactive sources: ^{60}Co and ^{137}Cs were used for calibration.

Figure 1. PARIS cluster. Marked are numbers of the detectors.

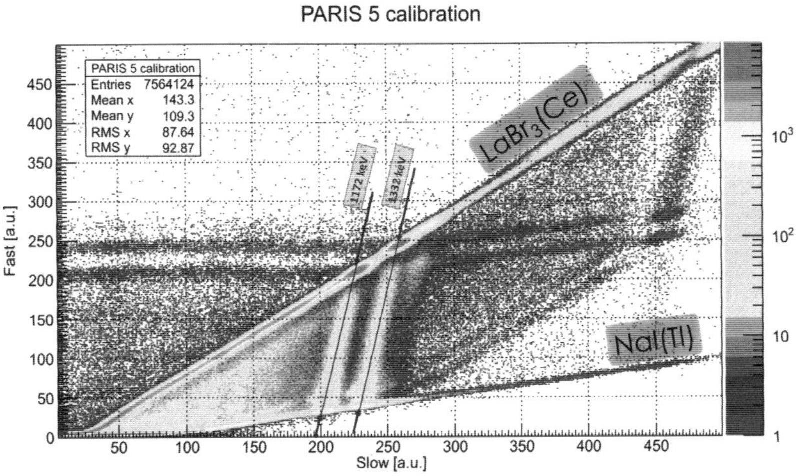

Figure 2. Fast vs. Slow matrix of one of the *phoswich* detectors from the calibration run with ^{60}Co and ^{137}Cs sources. Marked are lines of events understood as full energy deposition in one material only and parallel lines of the events corresponding to the same primary γ-ray energy.

3. Internal add-back

As was mentioned above, the signals from $LaBr_3$(Ce) and NaI(Tl) part of the *phoswich* detector differ by their rise time of typically 23 ns for $LaBr_3$(Ce) and 213 ns for NaI(Tl)[8]. Additionally $LaBr_3$(Ce) light yield is twice that of the one for NaI(Tl) which provides higher amplitude and bigger charge collected in the impulse for the same initial γ-ray energy. The process of separate, proper treatment of those signals and later building one common energy spectrum is a so-called *internal add-back*. The method would vary depending on the electronic set-up. In case of data from the BaFPro module it is possible to handle the fast and slow information as two versors and use a simple linear transformation of the basis they form.

There are two transformations of interest focusing on different aspects of Fast vs. Slow matrix. As one can see in Fig. 2, diverse events (pure deposition in one of the materials or scattered) with the same initial γ-ray energy form a line and lines corresponding to different γ-ray energies are parallel. Thus, energy spectrum can be simply obtained by rotation of a matrix by such an angle that those lines are parallel to Y-axis (Fig. 3). The energy spectrum can be built by taking into account only an X-component in the new basis.

The second approach uses the fact that the events with full energy deposition in only one material of the *phoswich* form a line. By the orthogonalisation process it is possible to change the basis to one where the energy deposited in $LaBr_3$(Ce) and energy deposited in NaI(Tl) part of the detector are used as versors (Fig. 4). This way the information of the place of the signal origin is conserved, however the number of needed parameters is larger resulting, due to uncertainties, in more unstable method. The energy spectrum is obtained by adding values of the two components.

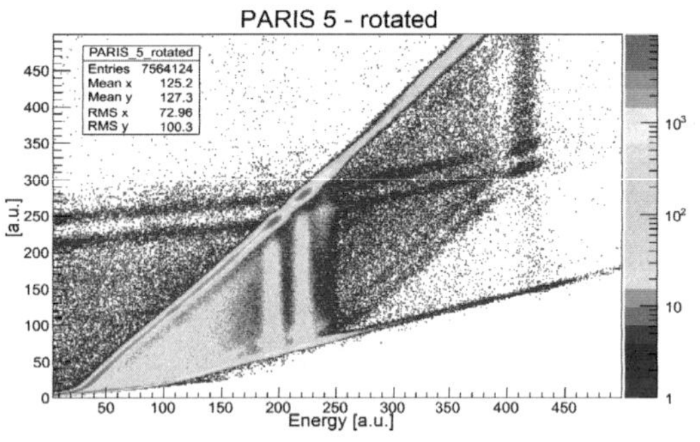

Figure 3. View of a rotated Fast vs. Slow matrix. Lines corresponding to events with the same primary γ-ray energy are parallel to the Y-axis.

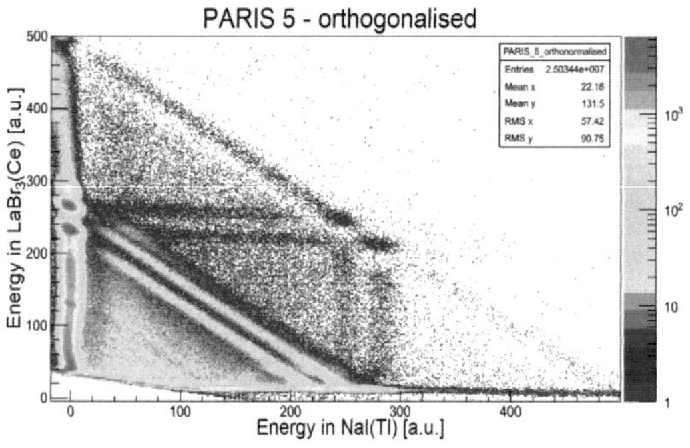

Figure 4. View of an orthogonalised Fast vs. Slow matrix. The angle of the lines corresponding to events with the same primary γ-ray energy is 135° in regard to the X-axis.

Unfortunately, the non-linearities of the data acquisition system for the low γ-ray energies exist. The angle between the lines of $LaBr_3$(Ce) and NaI(Tl) events is acute, thus the orthogonalisation stretches the plane between them, making the non-linearities especially visible in Fig. 4. Because of them the first approach, rotation transformation, proved to perform better than orthogonalisation and therefore became a method of choice for data from this experiment.

4. External add-back

For high energy γ-rays the primary interaction with the matter is via pair production. Hence, the creation of two secondary 511 keV photons is common, which in turn can escape from the detector creating single- and double-escape peaks in the energy spectra. The main advantage of the cluster configuration is a possibility to add the scattered γ-rays together and fill the original spectrum with the whole detected energy of the photon. This process is a so-called *external add-back*.

The employed algorithm is based on recurrence and is performed on calibrated data. When the detector provides a signal for an event, the surrounding detectors are also checked. All the energies are added together and fill the spectrum of the detector in which most of the energy was deposited.

In this analysis the external add-back procedure was performed after the internal add-back. Figure 5 illustrates the performance of the method showing spectra gated with multiplicities of photons detected in neighbourhood of the prime detector in one event. For higher multiplicities the ratio of single- and double-escape peaks to full absorption peak is getting smaller.

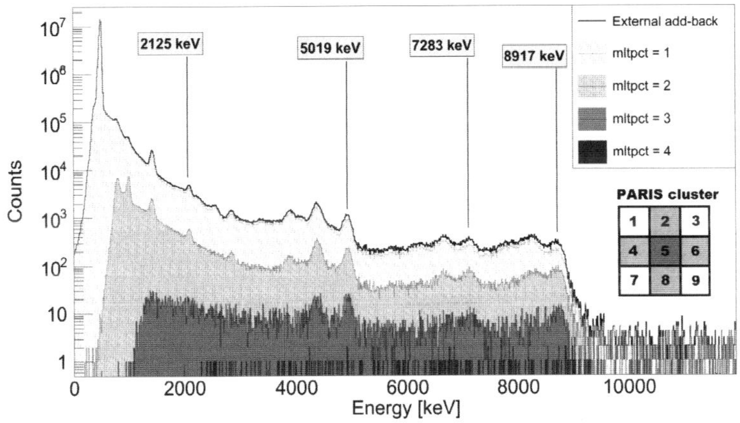

Figure 5. Reconstructed by external add-back procedure in-beam spectrum of γ-rays detected by the *phoswich* no. 5 and neighbouring detectors of the PARIS cluster (black line). Components of the external add-back spectrum corresponding to separate multiplicities are also shown. Marked are full absorption peaks. Increase in number of counts in the full absorption peak for 8917 keV line after external add-back is around 50%.

5. Energy resolutions

Energy resolutions were obtained with the use of ROOT framework Minuit library. Two fitting functions were used:

- $f(x) = p_1 + p_2 x + p_3 \exp\left(-0.5\left(\frac{x-p_4}{p_5}\right)^2\right)$ – a Gaussian function on linear background;
- $f(x) = p_1 + p_2 x + p_3 \exp\left(-0.5\left(\frac{x-p_4}{p_5}\right)^2\right) + p_6 \exp\left(-0.5\left(\frac{x-p_7}{p_8}\right)^2\right)$ – two Gaussian functions on linear background, where: p_1, p_2 – parameters of linear background; p_3, p_6 – parameters proportional to the peaks' integrals; p_4, p_7 – mean values of the peaks; p_5, p_8 – standard deviations of the peaks.

The second function was used for fitting peaks in the high-energy region where full absorption peaks significantly overlap with single-escape peaks. Those functions proved to stably reproduce standard deviation and centroids for various initial parameters. However, the rest of the fit parameters was more dependant on starting values.

Obtained preliminary values for the one of the detectors during subsequent steps and sum of all nine spectra after analysis (*full add-back*) are shown in Table 1. The values for pure

Table 1. Energy resolutions for one of the detectors during subsequent steps of data processing and for the whole cluster (full add-back).

Energy [keV]	Detector no. 5				Full add-back
	$LaBr_3$(Ce)	NaI(Tl)	Internal add-back	External add-back	
2124	2.59(31)%	3.40(85)%	3.33(15)%	3.28(19)%	3.26(05)%
5019	2.19(05)%	3.80(49)%	2.91(11)%	3.24(09)%	2.70(03)%
7283	1.81(16)%	3.17(1.0)%	2.31(29)%	2.64(25)%	2.60(09)%
8917	1.84(08)%	2.78(49)%	2.21(08)%	2.36(07)%	2.38(10)%

$LaBr_3$(Ce) and NaI(Tl) were evaluated with the use of gated on those events spectra from an initial, before any numerical procedure, Fast vs. Slow matrix. The number of events for high energy γ-rays in those spectra is small, resulting in high uncertainty for obtained values. As one can see the energy resolution of the *phoswich* is always worse than of the pure $LaBr_3$(Ce) part, yet it is better than of the pure NaI(Tl). The energy resolution slightly worsened for a spectrum after external add-back. It was possibly caused by the non-linearities of the data acquisition set-up. Evaluation of those non-linearities will be the next step of the data analysis.

6. Summary

The experiment held at the γELBE facility was an important step in evaluating the PARIS cluster performance. One of the noteworthy results of this test is the creation of the algorithms for the analysis of the data collected with the analogue BaFPro module. The evaluation of two algorithms for internal add-back showed, while using with BaFPro module, the better method to employ is the rotation transformation of a matrix. Comparing the energy spectra of $LaBr_3$(Ce) part only and the whole detector after internal and external add-back, a significant rise in full absorption peak height on the cost of slightly worse energy resolution was observed. This outcome, although expected, is promising.

Acknowledgements

This research was supported by the Polish Ministry of Science and Higher Education grants no. 2011/03/B/ST2/01894 and no. 2013/08/M/ST2/00591 and by the National Centre for Research and Development grant no. NCBiR/NUPNET/1/2011/w.

References

[1] Maj A *et al.* The PARIS Project. *Acta Phys. Pol B*, **40**:565–75, 2009.
[2] Beene J R *et al.* Heavy-ion excitation and photon decay of giant resonances in ^{208}Pb. *Phys. Rev. C*, **39**:1307–19, 1989.
[3] Shevchenko A *et al.* Fine structure in the energy region of the Isoscalar Giant Quadrupole Resonance: Characteristic scales from a wavelet analysis. *Phys . Rev. Lett.*, **93**(122501-1), 2004.
[4] Savran D *et al.* Nature of the pygmy dipole resonance in ^{140}Ce studied in (α, α/γ) experiments. *Phys. Rev. Lett.*, **97**(172502), 2006.
[5] Crespi F C L *et al.* Isospin character of low-lying Pygmy Dipole States in ^{208}Pb via inelastic scattering of ^{17}O ions. *Phys . Rev. Lett.*, **113**(12501), 2014.
[6] Schwengner R *et al.* The photon-scattering facility at the superconducting electron accelerator ELBE. *Nucl. Instrum. Meth. A*, **555**:211–19, 2005.
[7] Boiano C *et al.* Pulse shape results of $LaBr_3$ and BaF_2 scintillator obtained with a 16 ch. fast analog stretcher module. In *NSS/MIC*, pages 268–70. IEEE, 2010.
[8] Zieblinski M *et al.* Testing of the PARIS LaBr3-NaI phoswich detector with high energy gamma-rays. *Acta Phys. Pol B*, **44**:651–56, 2013.

Phototube non-linearity correction technique

S Riboldi[1-2], N Blasi[1], S Brambilla[1], F Camera[1-2], A Giaz[1] and B Million[1]

[1]Istituto Nazionale di Fisica Nucleare INFN, Sezione di Milano, Via Celoria 16, 20133 Milano (Mi), Italy.
[2]Università degli studi di Milano, Via Celoria 16, 20133 Milano (Mi), Italy.

E-mail: stefano.riboldi@unimi.it

Abstract. Scintillation light is often detected by photo-multiplier tube (PMT) technology. PMTs are however intrinsically non linear devices, especially when operated with high light yield scintillators and high input photon flux. Many physical effects (e.g. inter-dynode field variation, photocathode resistivity, etc.) can spoil the ideal PMT behavior in terms of gain, ending up in what are addressed as the under-linearity and over-linearity effects.

Established techniques implemented in the PMT base (e.g. increasing bleeding current, active voltage divider, etc.) can mitigate these effects, but given the unavoidable spread in manufacturing and materials, it turns out that, with respect to linearity at the percent level, every PMT sample is a story of its own.

The residual non linearity is usually accounted for with polynomial correction of the spectrum energy scale, starting from the position of a few known energy peaks of calibration sources, but uncertainly remains in between of calibration peaks. We propose to retrieve the calibration information from the entire energy spectrum and not only the position of full energy peaks (FEP), by means of an automatic procedure that also takes into account the quality (signal/noise ratio) of the information about the non-linearity extracted from the various regions of the spectrum.

1. Introduction

Radiation detectors based on recently developed scintillators with very high light yield and fast light emission (e.g. $LaBr_3$:Ce, with 63 photons/keV and 18 ns decay time [1]) may easily suffer from non-linearity effects of the associated photo-multiplier tube (PMT) devices, especially in case of high energy γ-ray/particle detection. In principle, also the non-linearity of the scintillator itself should be taken into account, but, particularly for $LaBr_3$:Ce and the energy range considered in this paper, that is not the dominant factor. Non negligible PMT under-linearity and over-linearity effects often emerge as the result of many interacting physical effects (e.g. inter-dynode field variation, photocathode resistivity etc.), which often can be mitigated, but not completely eliminated, by established techniques implemented in the PMT base (e.g. increased bleeding current, active voltage divider, etc.).

The residual non-linearity (e.g. of the order of a few percents and different from sample to sample of PMT) is usually accounted for with polynomial calibration of the spectrum energy scale based on the expected position of the energy peaks of a few calibration sources, but uncertainly still remains in between of the calibration peaks.

With this in mind, we start to introduce in section 2.1 an automatic algorithm that improves the signal-to-noise (S/N) ratio of the experimental energy spectra and increases the effectiveness of the new technique, proposed in section 2.2, that retrieves the calibration information from the entire energy spectrum and not only from the position of the known full energy peaks (FEP). In section 3 the effectiveness of the proposed methods is demonstrated using experimentally acquired energy spectra.

2. Methods

The experimental set-up used for acquisition of all the energy spectra is schematically represented in Fig. 1: a 3"x3" LaBr3:Ce doped scintillator (by Saint Gobain) is coupled to a Hamamatsu R10233-100-SEL PMT equipped with the LaBrVD PMT base developed in Milano, while a Lecroy HRO 12 bits oscilloscope sends the digitized PMT waveforms the a PC running the MATLAB software. PMT power supply is adjustable, in order to easily characterize the non-linearity as a function of the applied high voltage level.

Thanks to the new proposed method for non-linearity correction, which takes into account the entire shape of the spectrum, only a few calibration sources were required, as shown in Fig. 2, i.e. a ^{137}Cs source, an AmBe-Ni and the internal radioactivity of the LaBr scintillator,

2.1. Algorithm for signal-to-noise improvement

In order to improve the signal ("true counts") vs noise ("statistical fluctuations") ratio of the energy spectra, we propose a method based on a few steps: i) the energy axis E is rescaled to get uniform peaks FWHM (e.g. according to the square root of energy); ii) the counts axis C is rescaled to get comparable peak height (e.g. through to the logarithm function); iii) the resulting function C(E) is decomposed by Fourier techniques and synthesized back, neglecting the high-order harmonics, so that the intrinsic absolute energy resolution of the detector at low energy is preserved and at the same time the S/N ratio for high energy peaks for only few distributed counts is improved (see Fig. 3, 4, 5).

2.2. Algorithm for non-linearity correction

While ideal systems should have a perfectly proportional relationship between the physical energy of the detected events and the amplitude of the correspondent digitized waveforms, this is almost never the case with real detectors, especially if they are based on PMT technology.

Indeed, also in case of $LaBr_3$:Ce detectors the predominant source of non proportionality can be identified with the associated PMT device, apart from the tens-of-keV range of energy.

The non-linearity characteristic of a given combination of PMT and HV power supply level can be easily estimated by comparing two energy spectra obtained in the very same experimental condition apart from the two HV levels (see Fig. 6). If one energy spectrum is acquired with low-enough PMT gain so that all non-linearity effects are practically negligible, although resolution at low energy may be clearly compromised, it is always possible to characterize the PMT non-linearity at any given high voltage level by simply comparing the associated non-linear spectrum with the linear one.

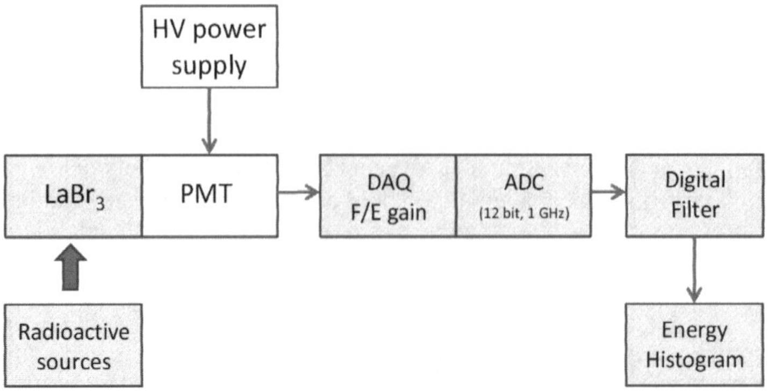

Figure 1. The experimental set-up used for the acquisition of the energy spectra. The adjustable gain of the DAQ F/E signal allows to fit the various amplitude of the PMT output signal, for different HV levels, to the fixed ADC dynamic range.

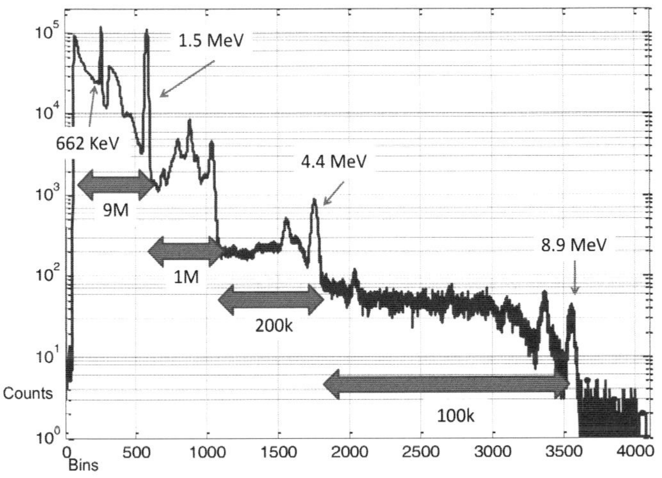

Figure 2. Reference energy spectrum (10 millions counts, acquired at 500 cps rate) ranging from 100 keV up to 9 MeV, with a few recognizable peaks from two calibration sources: ^{137}Cs (662 leV) and AmBe-Ni, plus $LaBr_3$ intrinsic radioactivity.
As high energy events are detected with lower efficiency, the S/N ratio in the right side is usually considerably lower.

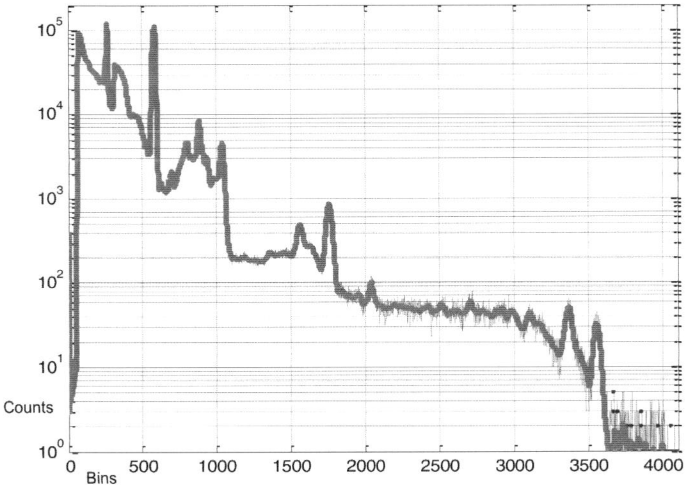

Figure 3. The energy spectrum of Fig. 1 (blue line) superimposed with a new version of the same spectrum with improved S/N ratio (red line), obtained by applying the procedure introduced in Sec. 2.2. Magnified regions of the two spectra are shown in Fig. 3 and Fig. 4.
The new energy spectrum, if required, can be easily re-sampled according also to different bin scales.

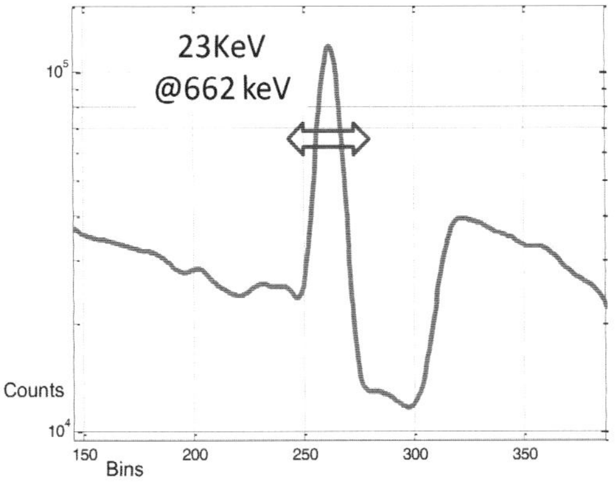

Figure 4. ^{137}Cs energy peak (662 keV) and surrounding region of the Fig. 3 spectra, showing total coincidence between the original and the improved S/N ratio spectra.

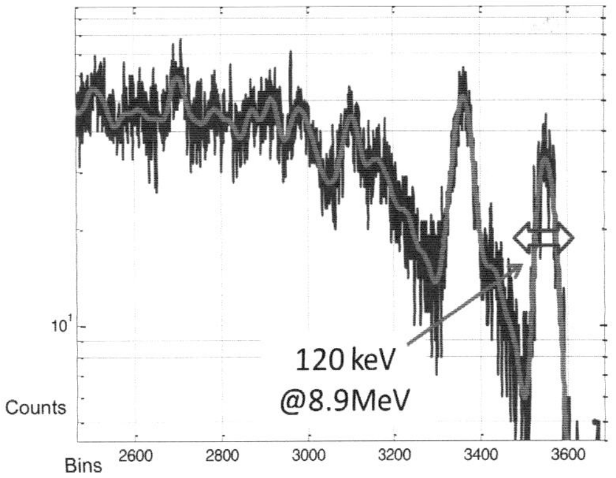

Figure 5. AmBe(Ni) full energy peak (8.9 MeV) with first escape peak and surrounding region of the Fig. 3 spectra, clearly showing the effectiveness of S/N improvement.

The final, slowly varying non-linearity correction function is determined by applying a Least Mean Squares (LMS) technique that takes into account: i) the non-linear spectrum shape; ii) the "shift" information, e.g. determined as in Fig. 7 and iii) the S/N quality of the shift information, i.e. higher in correspondence of narrow peaks (like the one in Fig. 4) and lower around nearly flat portions of the spectrum (like the 2300-3000 bin region of Fig. 3). An example of such a function is shown in Fig. 9, in the form of: bin(e(bin)+1).

Practically speaking, the entire procedure can be thought of as the search for a correction function that locally "stretches or shrinks" the ADC bin scale of the non-linear energy spectrum, so to make its shape as much as possible identical to the shape of the reference linear spectrum.

A more detailed description of the two methods will be readily published in a forthcoming technical paper.

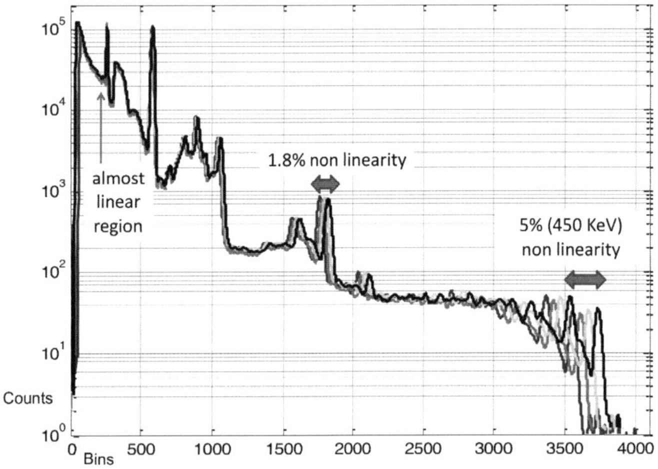

Figure 6. These 4 energy spectra have all been acquired using Fig. 1 set-up but different HV power supply level. While the blue spectrum (710 V, 15 mV@662 keV) is practically linear, significant over-linearity appears for higher HV values (e.g. 810 V for the red spectrum, with 2x signal amplitude compared to the blue one; 900 V and 4x amplitude for the green one; 975 V and 6x amplitude for the black one). Monotonic increase of non-linearity is evident for this PMT.

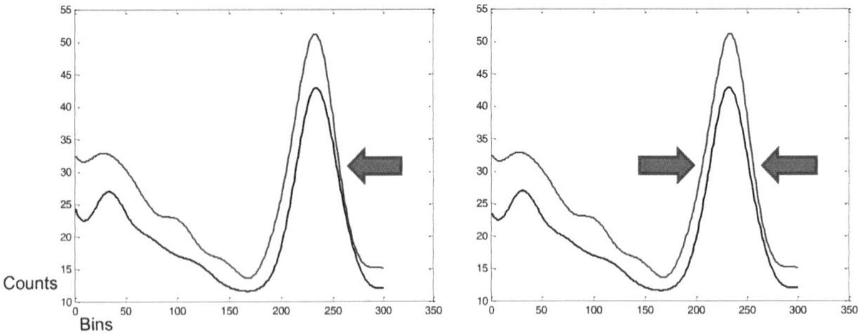

Figure 7. As an example of part of the procedure described in Sect. 2.2, the black peak from the non-linear spectrum (left panel) is left shifted and aligned with the blue one from the linear spectrum (as in right panel), thus maximizing the local cross-correlation function of the two spectra.

3. Experimental validation

As an applicative example of the proposed method, the red energy spectrum of Fig. 8 (acquired at 810 V) has been compared against the blue energy spectrum (acquired at 710 V and practically linear, but with degraded energy resolution at low energy). After estimating the non-linearity function (e.g. in the form of bin_L=bin*(1+e(bin) as the one represented in Fig. 9), the non-linearity of the red spectrum has been corrected and the two energy spectra have been superimposed in Fig. 10. Because the non-linearity correction introduces non-uniform binning in the energy scale, the resulting spectrum has been also uniformly re-sampled at 1 keV/bin, obtaining the final black energy spectrum in Fig. 10. Once the detector resolution as a function of energy is supplied to the algorithm, the entire procedure is fast and runs in a completely unattended way.

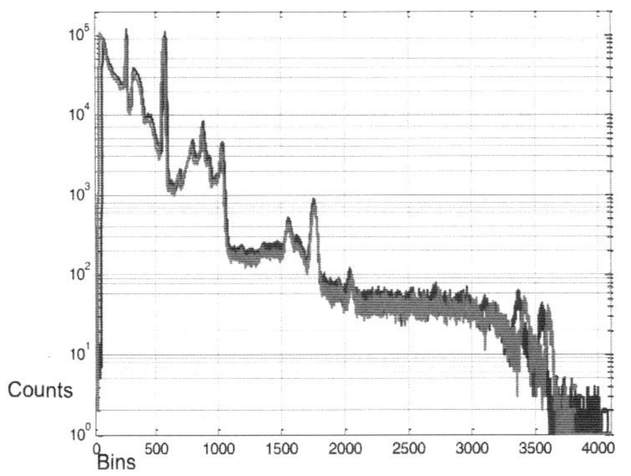

Figure 8. The two energy spectra have been acquired in the same experimental conditions, apart from the HV level. The blue spectrum at 710 V is practically linear, while the red spectrum at 810 V shows significant over-linearity.

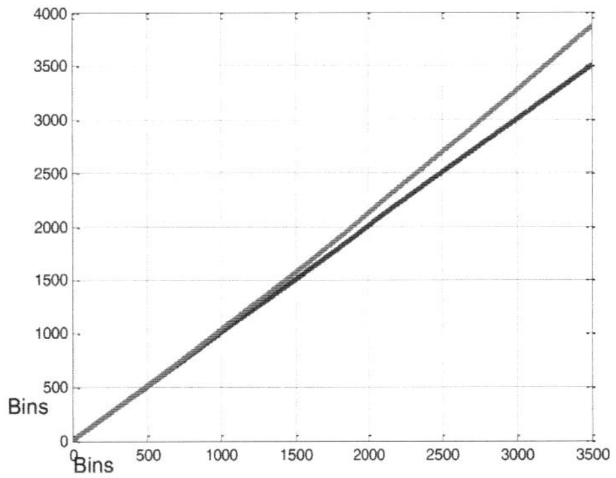

Figure 9. The Ideal bin_L=F(bin) function of a perfectly linear spectrum (blue) and the equivalent function (red), estimated using the proposed method, for a spectrum with significant 8% over-linearity at maximum energy.

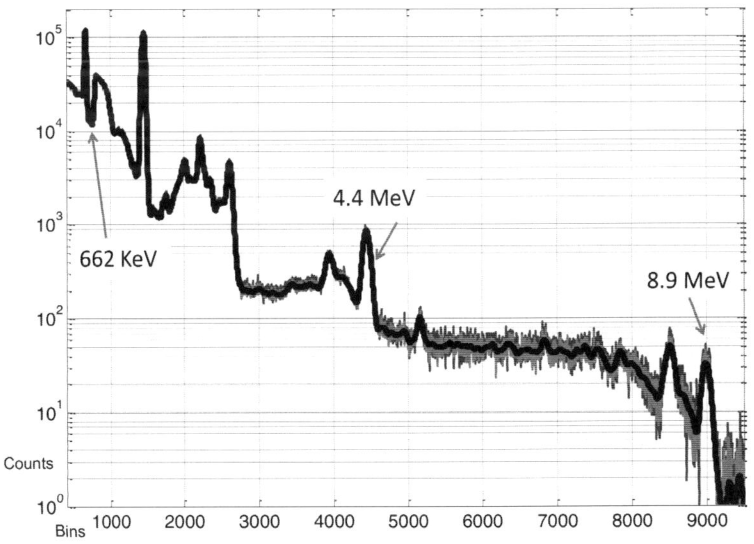

Figure 10. The effectiveness of the proposed method for non-linearity correction is shown here: the blue (linear) and red (linearized) spectra of Fig. 8 are now superimposed, while the black spectrum, derived from the red one after 1 keV uniform binning using the algorithm for S/N improvement.

4. Conclusions

A new method for PMT non-linearity estimation and correction has been developed, that takes into account the entire shape of the energy spectrum and not only the few peak positions of the calibration sources. Especially in case of $LaBr_3$ scintillator, this method can thus take advantage also of the detector significant internal radioactivity, to provide a quick and effective procedure for non-linearity correction that can also run unattended. Improvement of S/N ratio of the high energy regions of the spectrum is also obtained by a new proposed method based on Fourier signal decomposition. A more detailed analysis of the two proposed algorithms will be given in a forthcoming technical paper.

References

[1] Giaz A. *et al.*, 2013, *Nucl. Instr. and Met. A*, 729, 910-921.

SAMPIC: a readout chip for fast timing detectors in particle physics and medical imaging

Christophe Royon
IRFU/Service de Physique des Particules, CEA/Saclay, 91191 Gif-sur-Yvette cedex, France
E-mail: christophe.royon@cern.ch

Abstract. We describe the new fast timing readout chip SAMPIC developed in CEA Saclay and in LAL Orsay (France) as well as the results of differents tests performed using that chip.

1. Introduction: Timing measurements in particle physics and in medical imaging

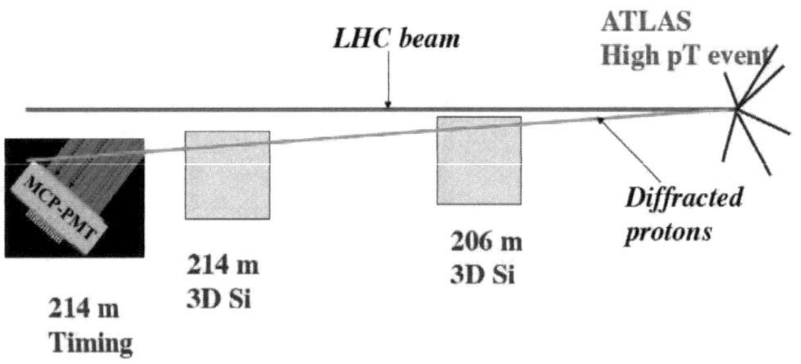

Figure 1. Scheme of the AFP proton detector in ATLAS. The same detector is implemented on the other side of ATLAS. A similar detector is installed in CMS/TOTEM.

At the Large Hadron Collider (LHC) at CERN, the most energetic proton-proton collider in the world with a center-of-mass energy of 14 TeV, there are special classes of events where protons are found to be intact after collisions. These events are called "diffractive" in the case of gluon exchanges. They can originate from photon exchanges as well. The physics motivation is a better understanding of diffraction in terms of QCD [1] and the search for beyond standard model physics such as the existence of extra-dimensions in the universe via anomalous couplings between photons, W and Z bosons [2]. The intact protons scattered at small angles can be measured in dedicated detectors, hosted in roman pots, located close to the beam and far away from the main central ATLAS or CMS detectors. In order to measure rare events at the LHC, the luminosity (or in other words the number of interactions per second) has to be as large as possible. In order to achieve this goal, the number of interactions per bunch crossing can be very large, up to 40-70 during the LHC running of 2015-2017 as an example. The projects to measure intact protons at high luminosity in the ATLAS and CMS/TOTEM experiments are respectively called AFP (ATLAS Forward Physics) and CT-PPS (CMS/TOTEM-Precision Proton Spectrometer) [3]. Timing measurements are crucial at the LHC in order to determine if the intact protons originate from the main hard interaction or from secondary ones (pile up). Measuring the proton time-of-flight with a typical precision of 10 ps allows constraining the protons to originate from the main interaction point of the event (hard interaction) with a precision of about 2.1 mm. For a pile up of 40 (which means about 40 interactions occuring in the same bunch crossing at the LHC), such a precision on time-of-flight measurements leads to a reduction in background of a factor of about 40 [4].

Timing measurements have also many applications in drone technology and in medical imaging as an example. The "holy grail" of medical imaging would be a PET detector with a 10 ps timing precision. With such an apparatus, image reconstruction is no longer necessary (the analysis can be performed online) since many fake coincidences can be suppressed, only attenuation corrections are needed, and real time image formation can be performed.

In order to achieve a 10 ps precision, many steps are needed going from the detector to the electronics and the readout software. In this article, we will concentrate on the achievement concerning the picosecond timing electronics that is currently being done in IRFU/SEDI Saclay and in LAL-Orsay [5, 6].

2. SAMPIC: SAMpler for PICosecond time pick-off

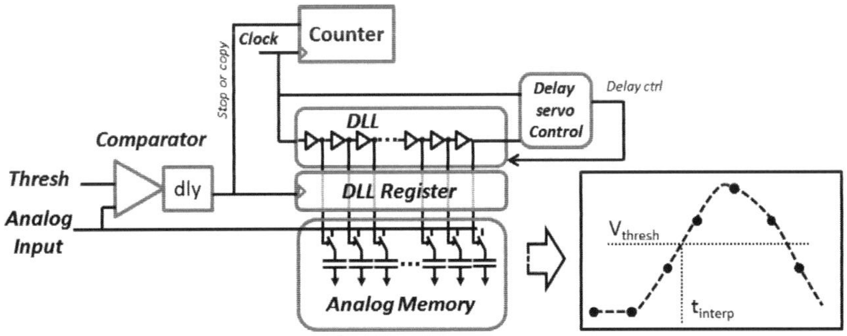

Figure 2. Scheme of the SAMPIC chip.

Before SAMPIC, the most performant Time to Digit Converters (TDCs) used digital counters and Delay Line Loops (DLLs). The timing resolution is limited by the DLL step and with most advanced Application Specific Integrated Circuit (ASICs), one gets a resolution of about 20 ps (new developments at CERN target 5 ps). The inconvenient is that a TDC needs a digital input signal: the analog input signal has to be transformed into a digital one with a discriminator which means that the timing resolution will be given by the quadratic sum of the discriminator and the TDC timing resolutions, thus leading to worse timing resolutions.

A new approach had to be developed using the principle of a wafeform based TDC. The idea is to acquire the full waveform shape of a detector signal in an ASIC dedicated to picosecond timing measurements. The input signal range has to be between 0.1 and 1. V with a fast rising time up to 1.ns in order to get the best possible performance of SAMPIC. The present version of the chip holds 16 channels (50 Ω terminated) with independent dead time. The possible trigger modes are either self triggered or triggered externally. Each channel includes an analog memory (64 cells) and recording is triggered by a discriminator. A Gray counter associated to DLLs allows assigning a time to the different samples and an ADC provides the conversion into a digital signal.

Figure 3. Picture of the SAMPIC chip and its acquisition board.

Three timing measurements with different precision are performed in SAMPIC. The time stamp Gray counter has a 6 ns step (it samples the reference clock), the DLL 150 ps (it defines a region of interest) and the waveform shape a few ps RMS after interpolation between the acquired points (they are acquired on a 64 step analog memory).

As we already mentioned, SAMPIC acquires the full waveform shape of a detector signal. The discriminator is used only for triggering, not for timing, and thus there is no jitter originating from the discriminator. All the information concerning the signal is kept in SAMPIC, and it is possible to use offline signal processing algorithms in order to improve the timing resolution. It can also be used to obtain other signal characteristics such as the deposited charge. In the present version, SAMPIC suffers an important dead time per channel due to the ADC conversion of about 1 μs. It will be reduced by about one order a magnitude in the next version of SAMPIC, using in particular the so called "ping-pong" method and analog buffering. Two SAMPIC chips can be hosted in a mezzanine board developed in LAL, Orsay, leading to a 32-channel system. The input into SAMPIC is sent via MCX connectors. SAMPIC can be read out using an USB-Ethernet-Optic fiber readout is also provided. A 5 V voltage power supply is the only element needed to run the mezzanine board and the readout software runs on Windows or Linux. A scheme of SAMPIC is given in Fig. 2 and a picture of SAMPIC together with its acquisition board in Fig. 3.

SAMPIC is quite cheap (about 10 Euros per channel) with respect to a few 1000s Euros for previous technology, which means that it can be used in large scale detectors such as PET for medical applications.

As a reference, a table giving the parameters of the SAMPIC chip is given in Fig. 4.

3. SAMPIC performance

3.1. Electronics tests

In this section, we describe the SAMPIC performance obtained from pure electronics tests. The maximum signal size is about 1.V, and after corrections, the average noise is quite low, of about 1 mV RMS (the noisiest cells being 1.5 mV RMS), which means a dynamic 10 bits RMS.

The SAMPIC cross talk was measured by sending a signal of 800 mV with a 300 ps rise time on one channel and reading out the neighbouring channels. The cross talk was found to be

		Unit
Technology	AMS CMOS 0.18μm	
Number of channels	16	
Power consumption	180 (1.8V supply)	mW
Discriminator noise	2	mV rms
SCA depth	64	Cells
Sampling Speed	<3-8.4 (10.2 for 8 channels only)	GSPS
Bandwidth	1.6	GHz
Range (Unipolar)	1	V
ADC resolution	8 to 11 (trade-off time/resolution)	bit
SCA noise	<1.3	mV rms
Dynamic range	9.6	Bit rms
Conversion time	0.2-1.6 (8bit-11bit)	μs
Readout time (can be probably be /2)	25 + 6.2/sample	ns
Time precision before correction	15	ps rms
Time precision after timing INL correction	< 5	ps rms

Figure 4. Parameters of the SAMPIC chip.

less than 1%. The quality of sampling was tested using a sinus wave signal, and the signal was perfectly reproduced without corrections at a sampling frequency of 10 Gigasamples per second. The sampling speed in SAMPIC is possible between 3 and 8.2 Gigasamples per second on 16 channels (up to 10 Gigasamples per second for 8 channels).

The timing resolution was studied by using two different channels of SAMPIC. The same signal was sent on both channels, one being delayed compared to the other one using a delay box or longer cables. The pulse had an amplitude of about 1.2 V, and we used the 6.4 Gigasamples per second configuration. The RMS of the time difference between the two signals as a function of delay is given in Fig. 5 using two different offline algorithms to reconstruct the time difference (CDF as constant fraction discriminator and CC as cross correlation using a linear or a spline interpolation between the different points measured by SAMPIC). The time resolution is quite flat as a function of the delay between the two signals and is about 5 ps, leading to a time resolution per channel of about 4 ps.

A similar study of the timing resolution versus the signal amplitude is shown in Fig. 6. The signal has to be above 450 mV in order to obtain the best timing resolution possible of about 4 ps.

3.2. Timing resolution using detectors

The second series of tests was performed by plugging SAMPIC into a real detector. We used a laser signal splitted in two, and going through two fast Si detectors [7]. The time difference between the two channels was measured using SAMPIC. The result is shown in Fig. 7 using the offline cross correlation algorithm. The time resolution is about 30 ps. It is of course dominated by the fast Si detectors, the resolution of SAMPIC being of the order of 4 ps. Additional studies are being performed in beam tests using diamond detectors leading to a time resolution of 80 to 90 ps [8].

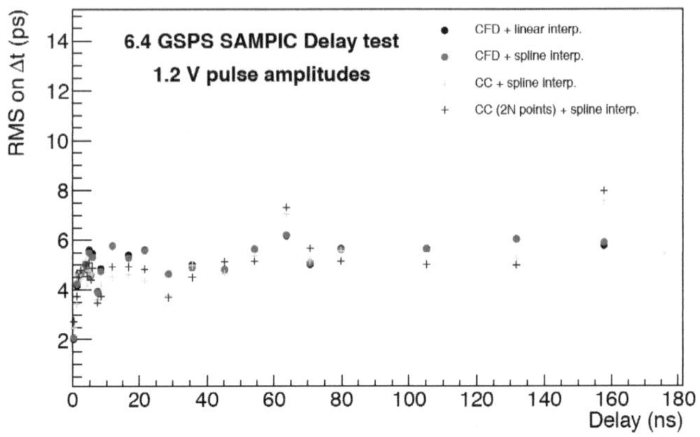

Figure 5. RMS on the time difference between two signals, one being delayed with respect to the other using two offline algorithms (Constant Fraction Discriminator (CFD) and cross correlation (CC)) using a linear or a spline interpolation.

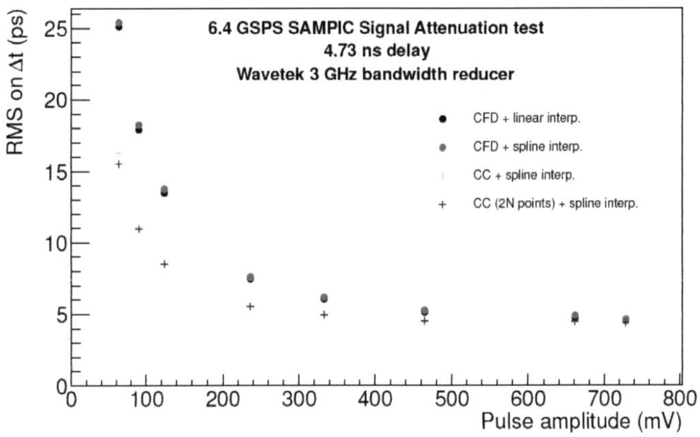

Figure 6. RMS on the time difference between two signals as a function of the signal amplitude using two offline algorithms (Constant Fraction Discriminator (CFD) and cross correlation (CC)) using a linear or a spline interpolation.

4. Conclusion

A self triggered timing chip demonstrator has been designed and characterised with 1.6 GHz bandwidth, up to 10 Gigasamples per second, low noise and of the order of 4 ps timing resolution. The chip is now ready and can be used for tests. Tests already started within the AFP, CT-PPS and CMS/TOTEM projects using quartz, diamond and Si detectors. Work is still going on in order to improve the chip concerning the DAQ system optimisation (firmware and software) and the improvement of the dead time using the "ping-pong" method. SAMPIC can now be used in many applications for tests in addition to particle physics for instance in medical imaging, drones, in detectors including many channels due to the low cost per channel.

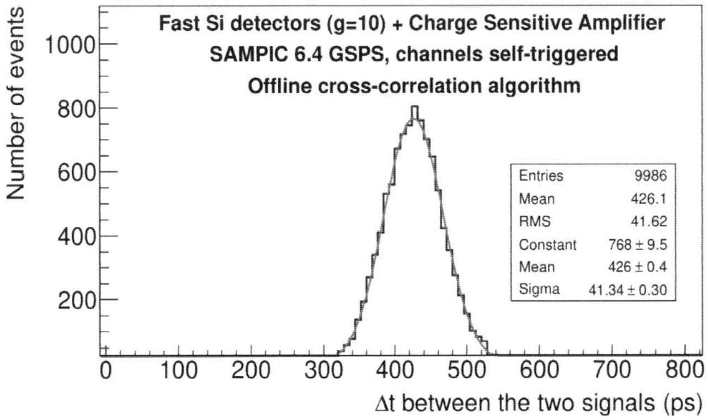

Figure 7. Time difference between two SAMPIC channels reading out Si detectors, a laser signal splitted in two going through the two Si detectors.

Acknowledgments

These results originate from a useful collaboration with D. Breton, V. de Caqueray, N. Cartiglia, E. Delagnes, H. Grabas, J. Maalmi, N. Minafra, M. Saimpert.

References

[1] C. Marquet, C. Royon, M. Saimpert, D. Werder, Phys. Rev. D**88** (2013) 7, 074029.

[2] S. Fichet, G. von Gersdorff, B. Lenzi, C. Royon, M. Saimpert, JHEP 1502 (2015) 165; S. Fichet, G. von Gersdorff, O. Kepka, B. Lenzi, C. Royon, M. Saimpert, Phys. Rev. D**89** (2014) 114004; E. Chapon, O. Kepka, C. Royon, Phys. Rev. **D81** (2010) 074003; O. Kepka and C. Royon, Phys. Rev. D **78** (2008) 073005; C. Royon, M. Saimpert, Int. J. Mod. Phys. A**29** (2014) 28, 1446016.

[3] N, Cartiglia, C. Royon, Int.J. Mod. Phys. A**29** (2014) 28, 1446017, A special issue on DPE/CEP, Editors: M. Albrow, V. Khoze, C. Royon; ATLAS Collab., CERN-LHCC-2011-012; TOTEM Collab., CERN-LHCC-2013-009, CERN-LHCC-2014-020, addendum CERN-LHCC-2014-024; CMS and TOTEM Collab., CERN-LHCC-2014-021; CERNLHCC-2014-026.

[4] C. Royon, M. Saimpert, O. Kepka, R. Zlebcik, Acta Physica Polonica B, Proceedings supplement, Vol. 7, Number 4 (2014) 735.

[5] E. Delagnes, H. Grabas, D. Breton, J. Maalmi, P. Rusquart, Acta Physica Polonica B, Proceedings supplement, Vol. 7, Number 4 (2014) 685.

[6] D. Breton, E. Delagnes, J. Maalmi, P. Rusquart, Acta Physica Polonica B, Proceedings supplement, Vol. 7, Number 4 (2014) 673.

[7] N. Cartiglia et al., Acta Physica Polonica B, Proceedings supplement, Vol. 7, Number 4 (2014) 657; N. Cartiglia et al., JINST **9** (2014) C02001.

[8] TOTEM Coll., beam tests performed at CERN/DESY, private communication.

Polycrystalline para-terphenyl scintillator adopted in a β^- detecting probe for radio-guided surgery

E Solfaroli Camillocci[1,2], F Bellini[1,2], V Bocci[2], F Collamati[2,3], E De Lucia[4], R Faccini[1,2], M Marafini[2,5], I Mattei[4,6], S Morganti[2], R Paramatti[2], V Patera[2,3], D Pinci[2], L Recchia[2], A Russomando[1,2,7], A Sarti[3,4], A Sciubba[2,3], M Senzacqua[1,2] and C Voena[2]

[1] Dip. Fisica, Sapienza Univ. di Roma, Rome, Italy; [2] INFN Sezione di Roma, Rome, Italy; [3] Dip. Scienze di Base e Applicate per l'Ingegneria, Sapienza Univ. di Roma, Rome, Italy; [4] Laboratori Nazionali di Frascati dell'INFN, Frascati (RM), Italy; [5] Museo Storico della Fisica e Centro Studi e Ricerche "E. Fermi", Rome, Italy; [6] Dipartimento di Fisica, Università Roma Tre, Rome, Italy; [7] Center for Life Nano Science@Sapienza, Istituto Italiano di Tecnologia, Rome, Italy.

E-mail: `elena.solfaroli@roma1.infn.it`

Abstract. A radio-guided surgery technique exploiting β^- emitters is under development. It aims at a higher target-to-background activity ratio implying both a smaller radiopharmaceutical activity and the possibility of extending the technique to cases with a large uptake of surrounding healthy organs. Such technique requires a dedicated intraoperative probe detecting β^- radiation. A first prototype has been developed relying on the low density and high light yield of the diphenylbutadiene doped para-therphenyl organic scintillator. The scintillation light produced in a cylindrical crystal, 5 mm in diameter and 3 mm in height, is guided to a photo-multiplier tube by optical fibres. The custom readout electronics is designed to optimize its usage in terms of feedback to the surgeon, portability and remote monitoring of the signal. Tests show that with a radiotracer activity comparable to those administered for diagnostic purposes the developed probe can detect a 0.1 ml cancerous residual of meningioma in a few seconds.

1. Introduction

A novel radio-guided surgery (RGS) technique using β^- radiation is being developed [1, 2]. The main advantage with respect to the traditional RGS with γ radiation [3] is the lower target-to-background activity ratio (TBR). β^- radiation indeed penetrates only a few millimetres of tissue with essentially no γ contamination being the *bremsstrahlung* contribution almost negligible. Low background from healthy tissue close to the lesion allows a smaller radiotracer activity to detect tumour remnants, and the possibility of extending the technique to cases with a significant uptake from healthy organs around the lesion.

In the past, the use of β^+ decaying tracers was proposed [4], since they are largely diffused in diagnostic applications (*Positron Emission Tomography*, PET). The emitted positrons have a limited penetration and their detection is local, but at rest annihilate with electrons producing γs with energy of 511 keV: the background persists and, actually, increases in energy becoming more penetrating. This approach has been studied in pre-clinical tests but it is not yet in use in clinical practice. The largest limitations range from the activity to be administered to achieve a quick probe response, to the complexity of the probe and to the exposure of the medical personnel.

On the contrary, pure β^- emitting tracers allow to develop a handy and compact probe which, being sensitive only to particles emitted locally and operating in a low background environment, provides a clearer delineation of margins of the lesioned tissue.

Meningioma brain tumour was identified as the first clinical case of interest because of its sensitivity to DOTATOC [5], a synthetic analogue of somatostatin, that can be marked with the pure β^- radionuclide ^{90}Y [6]. This isotope is suitable for β^- RGS because of its half-life (64 h), electron energy spectrum (endpoint: 2.28 MeV) and absence of γ emission. A study [7] of the potentiality of the proposed RGS technique has been performed on PET diagnostic exams of patients affected by meningioma after administration of ^{68}Ga-DOTATOC (the uptake for the tracer can be assumed independent from the linked radionuclide). The study shows that a TBR greater than 10 is observed in almost all cases, making the effectiveness of the technique very promising. To implement this RGS technique, we are developing an intraoperative probe detecting β^- decays based on para-terphenyl scintillator. In this article we describe the first prototype and its performance.

2. The β^- intraoperative probe

The first prototype has been designed having in mind the constraint coming from the meningioma clinical case. The instrument is optimized to provide a very small device with high sensitivity to low energy electrons, directional detection and a fast response (few seconds) compatible with the surgical activity. The required target spatial resolution is fixed by the typical volume (0.1 ml) of a surgical tumour residual in this clinical case.

The radiation sensitive element of the β^- probe is a small scintillator tip made of commercial poly-crystalline para-terphenyl doped to 0.1% in mass with diphenylbutadiene [9], manufactured by Detec-Europe. This material was adopted after a detailed study [10] due to its high light yield (~3 times larger than typical organic scintillators) and non-hygroscopic property, while its low density minimizes the sensitivity to photons. Different detector sizes have been tested (see 2.1), the best configuration is a cylinder of 5 mm in diameter and 3 mm in height. A black PVC ring with external diameter of 11 mm encloses the detector providing a shield against radiation coming from the sides. A 10 μm-thick aluminium front-end sheet covers the scintillator to ensure the light tightness. This assembly is mounted on top of an easy-to-handle aluminium cylindrical body (diameter 8 mm and length 14 cm).

In order to avoid risk of patient coming into contact with electrical devices, the scintillation light is guided by four 50 cm long optical fibres outside the probe to a Hamamatsu H10721-210 photo-multiplier tube (PMT). This photo-sensor module has an integrated high voltage power supply circuit requiring an input voltage as low as 5 V, making this device compatible with the surgical environment. The read-out electronics and the logic board are housed in a compact box that wireless connects to a remote monitor to display the counting rate.

2.1. Optimization of the scintillator design

Detection of non-penetrating low energy β particles does not require thick scintillators and the high light yield of the para-terphenyl largely compensates its short light attenuation length (about 5 mm [10]).

To maximize the sensitivity to β particles, the optimal thickness of the scintillator has been determined experimentally. A cylinder of para-terphenyl with diameter of 2.1 mm is connected to the PMT by one optical fibre and the thickness of this scintillator is gradually reduced starting from 4 mm until the counting rate on a point-like ^{90}Sr source with activity of 370 Bq (*Sr-source* in the following) is maximized. The ^{90}Sr is a long-lived β^- emitter (half-life: 29 years) with

Journal of Physics: Conference Series **620** (2015) 012009 doi:10.1088/1742-6596/620/1/012009

decay energy of 546 keV and no gamma radiation. It exists in secular equilibrium with its daughter isotope ^{90}Y resulting in the emission of two β^- particles. The radioactive source is sealed inside a 0.7 mm thick plastic disc. In Fig. 1 the counting rate stored on the *Sr-source* as a function of the para-terphenyl thickness is shown. The optimal point is found to be around 2 mm.

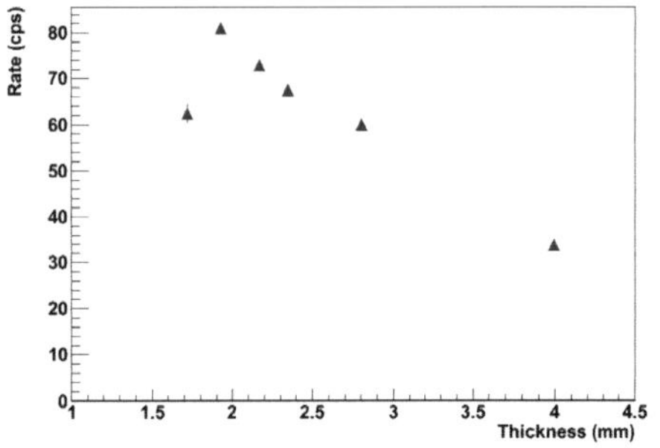

Figure 1. Probe counting rate as a function of the detector thickness.

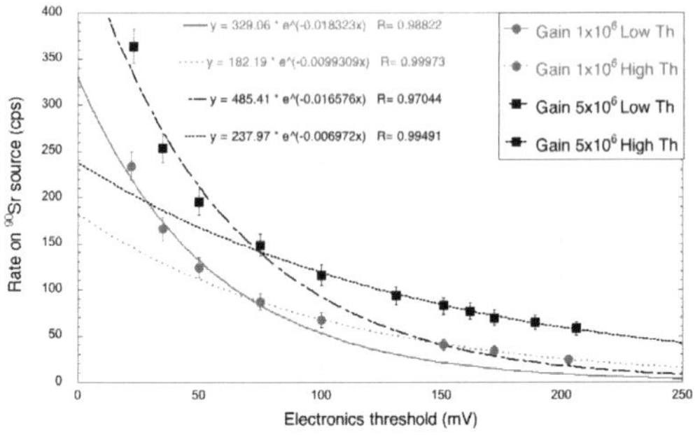

Figure 2. Probe counting rate vs electronics threshold. Two PMT gain settings are shown. In both cases after pulses are observed below 75 mV.

2.2. *The probe working point*

The probe working point is set by looking at the dark count and signal rates measured on the *Sr-source* by fixing the PMT gain value and varying the electronics threshold that trigger the pulse counter. The results are shown in Fig. 2 for two PMT gain settings: with a gain of $5{\cdot}10^6$ and a threshold of 80 mV the contribution of after pulses and dark current is below 0.2 cps with a signal rate of 140 cps.

The response stability to temperature variation in the range 10-30 °C has been measured separately for the scintillating crystal and the PMT. As shown in Fig. 3 the scintillator is almost insensitive to temperature variations whereas the PMT dark counts increase according to the Hamamatsu specification.

3. β^- detection sensitivity

The probe sensitivity in detecting electrons from β^- decays is measured with the *Sr-source* (nominal activity 370 Bq). The rate measured by the probe is $3.8{\cdot}10^5$ cps/MBq resulting in a detection efficiency of 40% of the β^- decay products of the chain ^{90}Sr/^{90}Y.

To study the capability of the probe to identify active spots, the detector is mounted on a motorized linear actuator ensuring position accuracy of 1.5 μm and horizontal scans are performed over point-like and extended sources. The scan over the point-like *Sr-source* (Fig. 4) shows how spatial resolution depends on the distance between the probe tip and the surface. The reconstructed profiles measured at distances ranging from 350 μm to 1.8 mm are fitted using a Gaussian distribution obtaining σ between 2.8 and 4.4 mm. The scan over the extended source (Fig. 5), instead, explores the discovery power of the probe in a more realistic simulation of a radio-labelled tumour residual. A cylindrical phantom with volume of 0.1 ml (diameter: 6.0 mm, height: 3.5 mm) is filled with a saline solution with ^{90}Y radionuclide of 1.4 kBq activity, comparable to those administered for diagnostic purposes. The profile reconstructed by the probe is obtained with an horizontal scan of 1.5 mm steps and 10 s per position and a distance

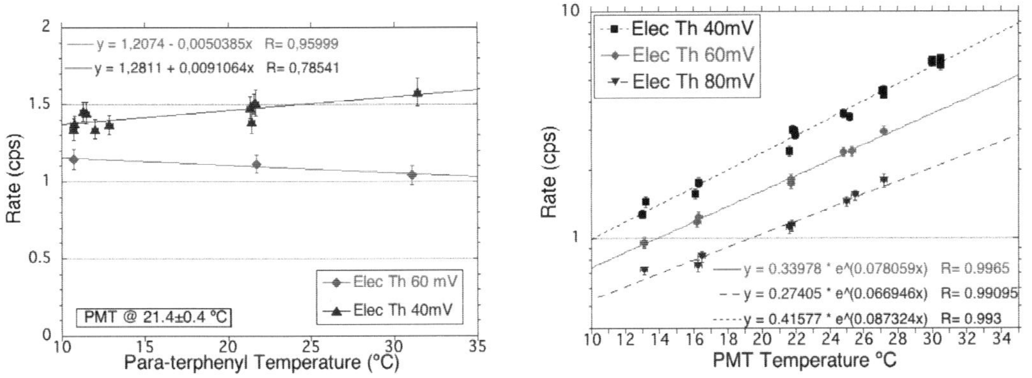

Figure 3. Test on stability of the probe electronic noise rate to temperature variations in the range 10-30 °C for the scintillator (left) and the PMT (right) separately. Configurations with different electronics thresholds are investigated.

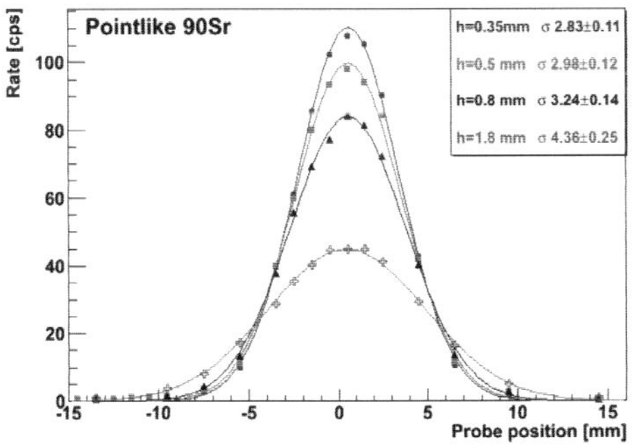

Figure 4. Profiles of the point-like *Sr-source* reconstructed by the probe at different distances from the source.

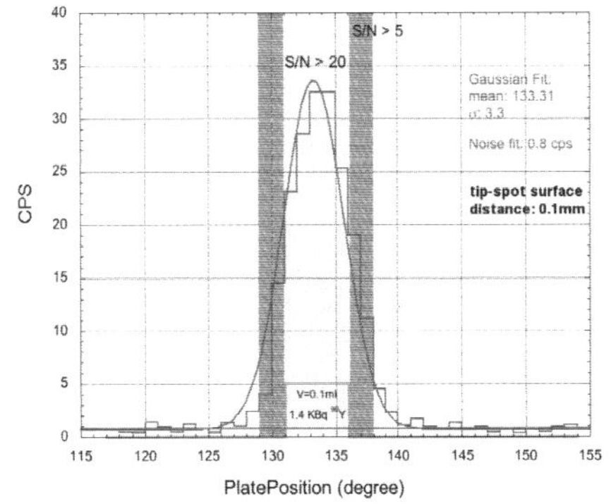

Figure 5. Lateral scan of a 1.4 kBq, 0.1 ml, ^{90}Y phantom. Gaussian and linear fits are for the signal and dark count rate respectively. See text for the band definition.

between the probe and the phantom surface of 100 μm. The probe is able to well identify the active spot with a signal discriminating threshold S/N>5, where S is the number of detected events and N is the number of dark counts, and the real phantom width is reconstructed with S/N>20.

4. Background rejection

The usage of pure β^- emitting tracers allows to operate with low radiation background, the residual being mainly due to photons coming from the *Bremsstrahlung* radiation of the electrons penetrating the tissue. The abundance of the expected *Bremsstrahlung* radiation as a function of the photon energy is shown in Fig. 6 (left) as computed with a simulation of a ^{90}Y source in water [11].

We measure the sensitivity of the probe to these photons using three point-like sources: ^{133}Ba emitting photons with energy ranging from 80 to 350 keV, ^{137}Cs with gamma emission at E_γ=662 keV and ^{60}Co with $E_{\gamma 1}$=1170 keV and $E_{\gamma 2}$=1330 keV. To avoid signal from electrons in

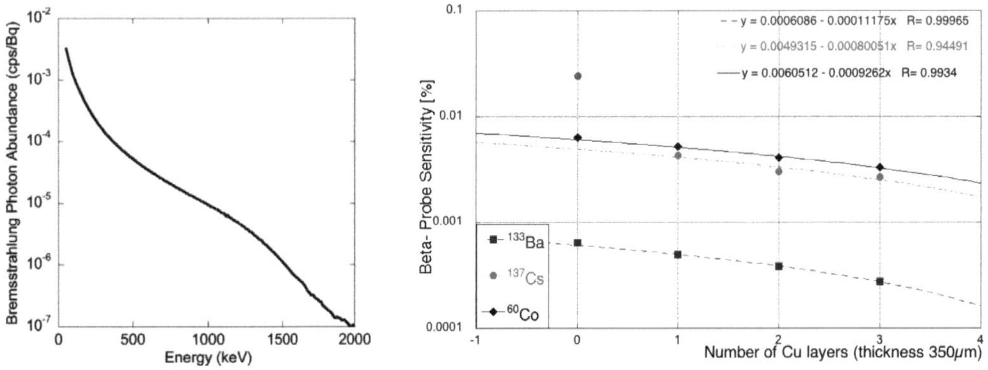

Figure 6. Left: Energy spectrum (50-2000 keV, 10 keV interval) of ^{90}Y *Bremsstrahlung* photons in water. **Right**: Probe sensitivity to photons emitted by three different sources: ^{60}Co $E_{\gamma 1}$=1170 keV $E_{\gamma 2}$=1330 keV, ^{137}Cs E_{γ}=662 keV, ^{133}Ba E_{γ} ranging from 80 to 350 keV. The measurements are repeated after insertions of copper layers with thickness of 350 μm between the source and the probe tip to absorb the electronic component of the Caesium emissions.

the case of Cs decays, three copper layers with 350 μm thickness are inserted in sequence between the source and the probe tip, and the measurements are repeated at each step. The counts as measured by the probe are shown in the Fig. 6 (right) for the three sources. Except for the first measure on the Cs source, introducing the copper absorbers implies a very small decrease in rate compatible with the attenuation in copper and the change in geometrical acceptance.

The sensitivity to the photons emitted by the ^{133}Ba source is below 10^{-5}, whereas for the γs from the ^{137}Cs and ^{60}Co sources the sensitivity is still below 10^{-4}. These measurements allow to conclude that the intraoperative β^- probe is not sensitive to the *Bremsstrahlung* photons and therefore the effectiveness of the RGS technique would not be affected by this background.

5. Conclusion

A prototype of the intraoperative probe for RGS exploiting β^- decays based on para-terphenyl scintillator has been developed and tested. The β^- detection efficiency in the energy range of the ^{90}Y emissions is >40%, while the sensitivity to the *Bremsstrahlung* photons is below 10^{-4} making this background radiation negligible. A test with a phantom simulating meningioma residual shows that with a radiotracer activity comparable to those administered for diagnostic purposes the probe is able to detect a 0.1 ml active spot in a few seconds.

References

[1] Solfaroli Camillocci E *et al* 2014 *Sci. Rep.* **4** 4401 DOI:10.1038/srep04401
[2] Patent PCT/IT2014/000025 deposited by Università degli studi di Roma "La Sapienza", Istituto Nazionale di Fisica Nucleare and Museo storico della fisica e centro studi e ricerche "E. Fermi".
[3] Povoski SP *et al* 2009 *World Journal of Surgical Oncology* 7-11
[4] Bogalhas F *et al* 2009 *Phys Med Biol* **54** 4439-53
[5] Bartolomei M *et al* 2009 *Eur J Nucl Med Mol Imaging* **36** 1407-16
[6] Heppeler A *et al* 1999 *Chem Eur J* **7** 1974-81
[7] Collamati F *et al* 2015 *J Nucl Med* **56** 1-6 DOI:10.2967/jnumed.114.145995
[8] Heute D *et al* 2010 *J Nucl Med* **51** 397-400
[9] Budakovsky SV, Galunov NZ, Grinyov BV, Kim JK, Kim YK, Tarasenko OA 2009 *Functional Materials* **16**, 1 86-91
[10] Angelone M *et al* 2014 *IEEE Transactions on Nuclear Science* **61**, 3 1483-87 DOI:10.1109/TNS.2014.2322106
[11] Rong X, Du Y, Frey EC 2012 *Phys Med Biol* **57** 371125 DOI:10.1088/0031-9155/57/12/3711

Thermal neutron detection by entrapping ^{6}LiF nanocrystals in siloxane scintillators

SM Carturan[1,2], T Marchi[2], G Maggioni[1,2], F Gramegna[2], M Degerlier[1,3], M Cinausero[2], M Dalla Palma[2,4], A Quaranta[2,4]

[1]University of Padova, Department of Physics and Astronomy, Padova, Italy.
[2]INFN - Laboratori Nazionali di Legnaro, Italy.
[3]Nevsehir Hacı Bektas Veli University, Science and Art Faculty Physics Department, Nevsehir Turkey.
[4]University of Trento, Department of Industrial Engineering, Trento Italy.

E-mail: carturan@lnl.infn.it, saramaria.carturan@unipd.it

Abstract. Exploiting the long experience in design and production of scintillating mixtures based on siloxane matrices with combinations of primary dye and waveshifter, a first set of ^{6}LiF loaded scintillator disks has been produced. The synthesis is herein described and reported, as well as preliminary results on their light response towards thermal neutrons. The preservation of transparency and mechanical integrity of the scintillator material is challenging when introducing the inorganic salt LiF which is a "foreign body" to the organic polysiloxane host matrix. Different strategies such as synthesis of nanoparticles and surface functionalization have been pursued to succeed in the entrapment of the neutron converter whilst maintaining moderate light output, optical transparency and flexibility of the base scintillator.

1. Introduction

The importance of revealing nuclear weapons concealed in trucks or boats at the borders, using Radiation Portal Monitors (RPM) has grown up, unfortunately, in the last years. Furthermore, there is a renewed interest in neutron detection both in nuclear physics, mainly related to the development of Radioactive Ion Beam facilities, and in applied physics related fields, such as materials analysis by neutron imaging and diffraction spectroscopy.

The use of ^{3}He for the detection of moderated neutrons with almost full rejection of γ-rays generated events has become prohibitive, due to the exceedingly high price/liter of this rare isotope, though the reliability of ^{3}He-based systems is still unrivalled. This has triggered a widespread and intensive research work aimed at ^{3}He replacement. Systems based on scintillators, both inorganic [1] and organic [2-5] have been recently proposed as neutron detectors, while compact and robust silicon based detectors where neutron converter thin films are used for signal generation have been recently ideated and proved their excellent performances [6]. Siloxane-based scintillators are, in principle, very simple and economic tools for particle detection. Their capability to detect thermal neutrons by loading with ortho-carborane has been demonstrated, but the limited solubility of the organoboron compound hampered their further development [7].

Other neutron converters have been tried as dopants for siloxane, such as Gd and Cd, but the toxicity of Cd itself and the intrinsic emission of γ-rays following neutron capture on Gd or Cd make those systems less attractive than ^{6}Li doped scintillators. In this case, the triton and alpha particles from the ^{6}Li reaction, as detailed by Knoll [8], can differently excite the material with respect to γ-rays, thereby enabling n-γ discrimination by pulse shape analysis (PSA).
In this preliminary work, we report on the production of a "standard" siloxane scintillator, added with a suitable combination of primary dye and wavelength shifter, where ^{6}LiF nanocrystals have been embedded. The choice of LiF as lithium bearing compound is dictated by several key features, such as minimal molecular volume, higher Li mass % content as compared to other Li salts and negligible hygroscopicity, which induces opacity and, in turn, severe light output loss. The critical issue to be addressed is the addition of ^{6}LiF to the siloxane matrix while preserving the light output. Therefore, the entrapment of nanoparticles, which are known to minimize light scattering effects, has been herein exploited for the first time. The synthesis of the nanocrystals of ^{6}LiF has been pursued by two different routes: the co-precipitation method in a water:ethanol solvent:co-solvent system [9] and the high temperature decomposition of trifluoroacetate lithium precursor (LiTFA) in oleic acid/octadecene environment [10].

2. Materials and methods

Polysiloxane based scintillators are obtained by Pt catalyzed addition reaction, as described previously [4,5]. In case of scintillators, one can add suitable dyes and neutron converter to the base resin, before vulcanization. In this case, 2,5-diphenyloxazole (PPO) was chosen as primary dye and Lumogen Violet (Basf, LV) as wavelength shifter.
As neutron absorber ^{6}Li was chosen because of its unique and clean decay channel consisting of a 2.73 MeV triton and 2.05 MeV alpha particle, as mentioned above. Among the various lithium compounds, lithium fluoride displays the lowest water absorption and, very recently, LiF nanoparticles synthesis has been reported [9,10]. In this work, ^{6}LiF nanoparticles were prepared using two different methods: co-precipitation and thermal decomposition of ^{6}LiTFA. In the first process, ^{6}Li chloride was prepared from hydrated ^{6}Li oxide, by treating the suitable amount in a pyrex capsule with hot concentrated HCl acid. Hence, the white salt was dissolved in a fixed volume of water:ethanol mixture to obtain a 0.2 M solution. Then it was slowly added to an already prepared solution of NH_4F in the same concentration and volume of solvent:co-solvent. The reaction was continued for about 1h, then the nanocrystals were collected by centrifugation.
In the other method, the preparation of ^{6}LiF with a shell of oleic acid, hereafter called ^{6}LiF OA, started from ^{6}LiTFA, obtained by treating hydrated ^{6}Li oxide with hot trifluoroacetic acid. Then, ^{6}LiTFA was mixed with oleic acid and octadecene (1:1 mixture) and the solution was heated under argon at 320°C for 1h. After cooling, excess ethanol was added and the nanocrystals with oleic acid functionalization were recovered by centrifugation.
Samples of polysiloxane with added ^{6}LiF were prepared with amounts of 0.5, 1 and 2 wt% of ^{6}Li by simple dissolution of oleic acid-capped particles, while mechanical blending was applied to disperse the nanoparticles from co-precipitation (water:ethanol 0:1). The samples with ^{6}LiF obtained from co-precipitation show several small undispersed grains, demonstrating the insolubility of the powders inside the siloxane matrix. On the other hand, ^{6}LiF OA containing samples, which are about 0.5 mm in thickness, appear as homogeneous, though as the amount of ^{6}LiF increases the optical clarity decreases remarkably. Samples with 0.5% of ^{6}Li were also obtained with different thickness (0.5, 1, 2 and 5mm). Only the samples containing ^{6}LiF OA have been tested with alpha and neutrons, since the other samples set cannot afford reproducible results.
Scintillation measurements were made by exciting the samples with a ^{241}Am α source, with a ^{60}Co γ-rays source and with a neutron AmBe source. Pulse height spectra were obtained by coupling the scintillator samples to an H6524 Hamamatsu PMT and wrapping them with aluminized Mylar on the front-face and Teflon tape on the side walls in order to maximize the light collection. In the case of neutron test, the AmBe source was shielded with polyethylene (6 cm) and lead (1 cm) bricks in order

to moderate neutrons and to block γ-rays. The yields were compared with that obtained from EJ-212 plastic scintillator (Eljen Technology Products, 1 inch diameter, 1 mm thickness) and lithium glass GS20 (Applied Scintillation Technologies, 1 inch diameter, 1 mm thickness) in the same experimental conditions.

3. Results and discussion

In figure 1 SEM images of the powders obtained by co-precipitation are shown. When equal amount of water and ethanol is used, micron sized cubes, highly regular with a quite narrow size distribution are produced. If only ethanol is used, the particles are much smaller and their size decreases down to some hundreds of nm.

Figure 1. SEM images of 6LiF crystallites obtained using different ratios water:ethanol (top panel 1:1, bottom panel 0:1).

In the case of LiTFA decomposition, SEM inspection evidences flower-like structures, as visible in figure 2. Oleic acid may promote aggregation of nanocrystals during growth at high temperature, with formation of different shapes and size in the order of some microns.

High-resolution X-ray diffraction analysis has been performed on powders (figure 3). All the samples show the typical XRD pattern of crystalline LiF (griceite). As for the particle size, the FWHM of the main peak (200) can be observed for a qualitative comparison. A slight peak broadening is visible for the case of co-precipitation synthesis as the amount of water decreases, indicating a smaller crystallites size. The peak width of ^{6}LiF OA sample is quite similar to the other cases and from the Scherrer formula, the crystallites size has been estimated to about 25 nm.

The samples with ^{6}LiF OA have been tested under alpha irradiation and the results are reported in figure 4 (left). As compared to EJ212, the light output reaches high values, up to 80% of EJ212, almost the same value of GS20. Sample thickness detrimentally affects the light collection, owing to reabsorption effects, as can be seen from light output decrease for increasing thickness. The energy resolution, estimated on the basis of the full width at half maximum (FWHM) of the peak, accounts for small deviation from this general behaviour. Nonetheless, in spite of its low transparency, the sample 5 mm thick still displays a moderately high light output (about 50% of EJ212).

Figure 2. SEM image of ^{6}LiF OA capped aggregate.

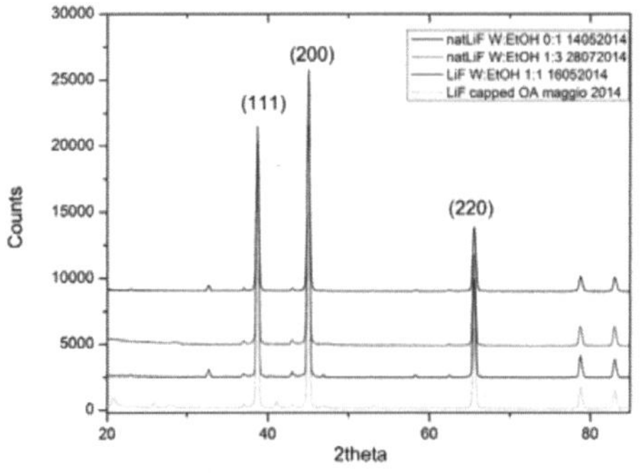

Figure 3. HR-XRD patterns from powders synthesized by the two different pathways.

Exposure to thermal neutrons allowed to record the spectra of figure 4 (right side), where the spectra from samples with 0.5% ^{6}Li and different thickness are reported and compared with GS20 standard. The samples spectra were collected with double gain with respect to GS20 as set on the amplifier. The response of GS20 appears as composed of more than one component (inset graph in figure 4, right panel). This can be ascribed to edge effects, where the produced ionizing particles, either alpha or triton, escape from the disk borders with incomplete energy deposition. However, the observed behaviour is still under discussion. The 0.5 mm thin samples obtained by adding higher amount of ^{6}LiF produced very low response to thermal neutrons, owing to the severe lack of transparency and low active volume for neutron capture. On the other hand, ^{6}LiF OA (0.5 wt%) dissolved in siloxane based scintillator with thickness 2 mm and 5 mm displayed a good response in light output, in spite of the quite low transparency of the latter (photos in figure 4).

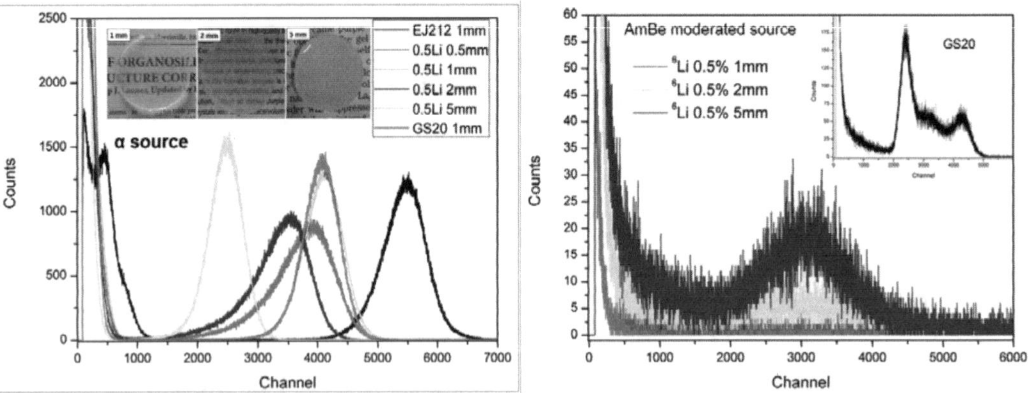

Figure 4. Pulse-height spectra collected by exposing the samples with 0.5 wt% 6Li and different thickness to alpha particles (left) and thermalized neutrons (right). Photos of the samples are reported in the inset on the left, while the neutron spectrum from GS20 is in the inset on the right, for the sake of clarity.

Neutron absorption within a scintillating material is described by the exponential attenuation of the impinging neutron beam, when passing through the material. The attenuation level depends on the absorbing nucleus concentration, its capture cross-section and the thickness of the absorbing layer. The neutron absorption length l is defined as the material thickness necessary to attenuate the original fluence of the impinging neutron beam down to a factor $1/e$ (i.e. about 63% absorption) [1]. In the case of GS20, which contains higher amount of ^{6}Li (1.6×10^{22} at/cm^{3}) and has higher density (2.5 g/cm^{3})

than our 0.5 wt% loaded samples (^{6}Li 5×10^{20} at/cm^3, density 1.03 g/cm^3) the absorption length is about 0.7 mm for thermal neutrons, while in the case of siloxane is about 20 mm. Therefore, it can be inferred that the higher thickness in siloxane based samples accounts for higher peak intensity and better signal-to-background ratio in the pulse-height spectrum, as visible in figure 4. On the other hand, the simple approach of increasing thickness to get still enhanced active volume for capture and, in turn, improved neutron sensitivity is not straightforward. In fact, the light pulse generated by the ionizing particles from the capture reaction (alpha and triton) must travel through the material until reaching the PMT window and self-absorption can severely hamper this process. Thicker samples should be prepared to get insight on the upper limit in thickness where light transport limitations counterbalance the larger capture volume.
Another interesting aspect is the γ-rays background, visible as a shoulder in the low channels region, which derives from the AmBe source not completely shielded. The thicker the sample the more significant becomes the contribution from Compton interactions. Nevertheless, the neutron signal is clearly well separated from the shoulder due to γ-rays background, thus indicating that n-γ separation is possible simply on the basis of pulse height discrimination.

4. Conclusions

^{6}LiF nanoparticles have been synthesized both by co-precipitation and ^{6}LiTFA thermal decomposition in oleic acid environment. In the first case, ^{6}LiF nanoparticles with variable size have been obtained controlling the ratio between solvent and co-solvent. However, the entrapment of the nanocrystals into the siloxane matrix produced non homogenous pellets, with visible granules and agglomerates. On the other hand, ^{6}LiF crystals synthesized with a cap layer of oleic acid proved to be compatible with siloxane matrices. Samples with different thickness and loadings of neutron absorber have been prepared, bearing moderate solubility and optical clarity. Test with neutrons from a moderated AmBe source demonstrated that, in spite of the low content of ^{6}Li, siloxane based scintillators can be used to detect thermal neutrons, as proved by the appearance of the capture peak in the pulse height spectrum, provided that the scintillator thickness is high enough to capture a good fraction of impinging thermal neutrons, but low enough to limit light loss by self-absorption.

Acknowledgments

Authors are very thankful to Mr Selvino Marigo and Mr Alessandro Minarello for their precious technical work. This work has been financially supported by INFN Commission IIIrd and Vth and by the Ministry of Education Education (Prin 2010-2011 -2010TPSCSP_005- *Developments of new detectors and analysis techniques for experiments with radioactive beams at the National Laboratories of INFN, with special reference to the SPES project*).

References

1. Van Eijk CWE, Bessiere A, Dorenbos P, 2004 *Nucl. Instrum. Meth.* A **529** 260.
2. Zaitseva N, Newby J, Hamel S, Carman L, Faust M, Lordi V, Cherepy N, Stoeffl W, Payne S, 2009 *Neutron detection with single crystal organic scintillators*, Proc. SPIE 7449.
3. Bell ZW, Brown GM, Ho CH, Sloop FV, 2003 *Organic scintillators for neutron detection*, Proc. SPIE 4784.
4. Carturan S et al., 2011 *Rad. Prot. Dosimetry* **143** 471.
5. Carturan SM, Marchi T, Fanchini E, De Vita R, Finocchiaro P, Pappalardo A., 2014 *Eur. Phys. J. Focus Point* **129** 212.
6. Barbagallo M et al., 2013 *Rev. Sci. Instrum.* **84** 033503.
7. Quaranta A et al., 2011 *J. of Non-Cryst. Solids* **357** 1921.
8. Knoll GF 1979 *Radiation Detection and Measurement* (John Wiley & Sons, Inc., New York), chapter 14 p 519.
9. Alharbi ND, Salah N, Habib SS, Alarfaj E, 2013 *J. Phys. D: Appl. Phys.* **46** 035305.
10. Du Y-P, Zhang Y-W, Sun L-D, Yan C-H, 2009 *Dalton Trans.* **40** 8574.

Thermal neutron scintillators using unenriched boron nitride and zinc sulfide

J E McMillan[1], A J Cole[1], A Kirby[1], E Marsden [2,4]

[1] Department of Physics and Astronomy, The University of Sheffield, Sheffield, South Yorkshire, S3 7RH, UK

[2] Tata Steel, Swinden Technology Centre, Moorgate, Rotherham, South Yorkshire, S60 3AR

E-mail: j.e.mcmillan@sheffield.ac.uk

Abstract. Thermal neutron detectors based on powdered zinc sulfide intimately mixed with a neutron capture compound have a history as long as scintillation technique itself. We show that using unenriched boron nitride powder, rather than the more commonly used enriched lithium fluoride, results in detection screens which produce less light but which are very considerably cheaper. Methods of fabricating large areas of this material are presented. The screens are intended for the production of large area low cost neutron detectors as a replacement for helium-3 proportional tubes.

1. Introduction

^{3}He proportional tubes were the industry standard neutron detectors over the period ~1970-2008. Recently, however, the international shortage of ^{3}He has led to the abandonment of these and an urgent search for alternative technologies [1].

Large numbers of neutron detectors are needed for Radiation Portal Monitors (RPMs) which are used at international borders and other locations to detect illegally shipped nuclear materials [2]. Identical monitors are employed in the metal recycling industry to detect orphan sources or other radioactive materials in scrap feeds before melting [3]. A further application is ensuring the security of sites where nuclear materials are used or stored.

As the expected fluxes are low, the detectors used must be as large as possible, square metres being typical. The detection efficiencies must be high, but also unambiguous, so good signal-to-noise ratios are desirable. Real-time signal discrimination is essential, as opposed to the computed post-processing encountered in many experimental physics environments. The detectors must be deployable, that is, reasonably robust and stable over many years operation in harsh environments. They must be manufacturable, transportable and installable with minimal health and safety implications and operable by unskilled staff. Lastly, because of the large areas required, there are constraints on price, which translate into a requirement to use readily available materials.

2. ^{6}LiF-ZnS scintillation detectors

Thermal neutron detectors based on powdered zinc sulfide intimately mixed with a neutron capture compound have a history as long as scintillation technique itself.

[4] Present address: Kromek Ltd, Netpark, Thomas Wright Way, Sedgefield, Durham, TS21 3FD, UK

Large volume thermal neutron detectors based on a mixture of isotopically enriched lithium fluoride and zinc sulfide scintillator powders held in a binder were demonstrated by Barton et al. [4, 5]. These used the scintillator mixture in layers, interleaved with acrylic wavelength shifting lightguides and polypropylene moderators. The lightguides were end-viewed by photomultipliers and neutron discrimination was provided by pulse counting discriminators similar to those described by Caines [6] and Davidson [7]. These detectors were originally built to study rare fission events and were operated in harsh environments in underground low-background sites.

The individual detectors had an active volume 90 × 14.4 × 14.4cm and a stack of eight detectors completely surrounding a ^{252}Cf source had a neutron detection efficiency of 37%. The thin sensitive layers and the pulse counting discrimination technique ensured that these detectors were totally insensitive to gammas or muons. They have provided robust stable operation over the intervening years, having been deployed in three different underground sites.

3. Neutron scintillation detectors for portal applications

A program to re-investigate this type of detector for use as the neutron sensitive component in RPMs has been undertaken.

Initially consideration was given to replacing the zinc sulfide. The first complaint about zinc sulfide is that it absorbs its own light and can only be used in thin layers. However, it is just this that gives the Barton design its immunity to gamma rays, as no gamma with typical nuclear decay energy will ever deposit sufficient energy in the thin layer to compare with the products of neutron capture on ^{6}Li. Zinc sulfide also has problems with long decay time and poor pulse height discrimination but these were largely dealt with by the pulse counting discriminators. Problems with long afterglow remain, but as high event rates are not expected, this is unlikely to be a problem. The fact remains that, despite half a century of research into inorganic scintillators, zinc sulfide remains the brightest low cost scintillator. The choice was made to continue with it.

^{6}LiF is a controlled material and is also increasingly expensive, so it was necessary to consider the alternatives. The only three capture reactions of interest are:

$$\begin{aligned} {}^3\mathrm{He} + \mathrm{n} &\rightarrow {}^3\mathrm{H} + {}^1\mathrm{H} + 0.764\mathrm{MeV} \\ {}^6\mathrm{Li} + \mathrm{n} &\rightarrow {}^4\mathrm{He} + {}^3\mathrm{H} + 4.8\mathrm{MeV} \\ {}^{10}\mathrm{B} + \mathrm{n} &\rightarrow {}^7\mathrm{Li} + {}^4\mathrm{He} + 2.3\mathrm{MeV} + 0.48\mathrm{MeV}(\gamma) \end{aligned}$$

Gadolinium, cadmium and europium are excluded, since their capture reactions involve emission and subsequent detection of gamma rays and, consequently, it is difficult to get good neutron/gamma discrimination with these elements. There are also issues of cost and toxicity.

Table 1 shows the natural abundances of the useful isotopes together with the thermal capture cross-sections for these reactions. The use of natural lithium is unattractive due to the low concentration of ^{6}Li. Natural boron on the other hand is attractive, with the 19.8% abundance

nuclide	% natural abundance	thermal cross-section
^{3}He	0.00014	5333
^{6}Li	7.50	940
^{10}B	19.80	3842

Table 1. Natural abundances and cross-sections in barns for useful capture reactions.

of the ^{10}B being offset by the very much higher capture cross-section. It is clear that the use of material which is not isotopically enriched is likely to result in considerable savings and that, even if the resulting detectors are slightly worse in performance, they may still be very attractive. If a detector can be demonstrated using unenriched boron, then it is clear that a considerable improvement can subsequently be obtained by switching to enriched material, but that this would also dramatically increase the cost.

The obvious disadvantage of the ^{10}B reaction is that the reaction products have much less energy than is the case in the ^{6}Li reaction. This will result in less light produced in the zinc sulfide and may require better designed optics. The other problem is that the products of the capture reactions have reduced range. Simulations using SRIM [8] indicated that the mean range of the ^{7}Li nucleus was 2.5μm while the alpha particle had a range of 4.3μm. This indicated that very fine grain size powdered boron compounds were needed.

Boron compounds mixed with zinc sulfide scintillator were demonstrated [9, 10] as neutron detectors at an early stage in the development of scintillation technique and much subsequent work was done, but this was largely abandoned when ^{3}He tubes became dominant. No attempt had been made to read it out using the Barton volume configuration [4], so all the reports of detectors using boron describe rather small detectors.

The ideal boron compound for this application is required to be completely chemically stable and transparent to the light emitted by the zinc sulfide. It should also be commercially available at low cost in controlled grain sizes. A boron compound which meets these requirements is boron nitride, BN, which is available in two allotropes; a hexagonal form and a cubic form. Hexagonal BN is a white slippery solid with a layered structure, physically similar to graphite. It is manufactured for lubricant and cosmetics applications.

Hexagonal BN has a density of 2100 kg m^{-3} and is birefringent with refractive indexes of 1.65 and 2.13. The pure material is highly transparent in the visible and near UV having essentially no optical absorption at wavelengths longer than 230nm [11]. The macroscopic neutron cross section of *BN is 38.93 cm^{-1}, compared to 57.53 cm^{-1} for ^{6}LiF, so the resulting detectors will be of comparable efficiency.

4. Optimization

The Barton detectors [4] were optimized for maximum efficiency and lowest background in a volume configuration. Security applications need to optimize effective area per unit cost:

$$\frac{\text{area} \times \text{efficiency}}{\text{cost}}$$

MCNPX [12] simulations indicated that in the original eight layer design, the four best layers contributed 76% of the efficiency of the detector. This suggested that the four worst layers can be redeployed at the side to double the area giving an effective area 1.52 times the original.

5. Scintillating layers

The layer production technology has been revised for ease of use and to reduce to reduce wastage. Originally a thermoset resin was used as a binder, but this was problematic in that it set rapidly and could not be reworked or easily cleaned from equipment. Experiments on replacing this with solvent soluble polymers were successful, a thick paint being made from ZnS and BN powders, polymer and solvent.

This is applied in a controlled thickness to a substrate of aluminized mylar. When the solvent evaporates, the scintillator layer remains firmly attached to the mylar and the result is sufficiently robust and flexible to withstand handling and even curving to quite a tight radius. Small pieces can be made by a doctor-blade spreading technique. To produce larger areas, a coating machine has been developed, which uses a peristaltic pump to apply the scintillator paint mixture to the mylar through a small nozzle. This can be seen in figure 1.

Figure 1. The coating machine

6. Wavelength shifting lightguides

With less light available from the boron reaction, than the lithium reaction, light collection becomes more important. The original BBQ loaded acrylic (Plexiglas GS2025) is no longer available. A number of alternative materials have been investigated. Some of the commercial wavelength shifting materials are based on polyvinyltoluene and are consequently intrinsic scintillators. Their use degrades the overall neutron/gamma discrimination of the current design. Success has been achieved with a commercial green fluorescent acrylic rod as can be seen in figure 2, however, supplies of this material may prove unreliable. Experiments have been made with dyed acrylic [13] and these are ongoing.

7. Current status

At present two experimental detectors using boron nitride are working. One is based on sheet wavelength shifter, while the other is based on cylindrical rods. They have the same gamma ray rejection as the original designs and similar detection efficiency to ^{3}He proportional tubes. Considerably more details can be found in Marsden's thesis [14]. More accurate efficiency measurements are in preparation and will be published shortly.

8. Acknowledgements

This work was supported by the Home Office Scientific Development Branch and subsequently by STFC CLASP award ST/K000179/1.

Figure 2. The detector based on cylindrical lightguides

References

[1] Kouzes R 2009 The ^{3}He supply problem Tech. Rep. PNNL-18388 Pacific Northwest National Laboratory, Richland, Washington, USA

[2] IAEA 2008 Illicit nuclear trafficking: collective experience and the way forward Tech. Rep. STI/PUB/1316 International Atomic Energy Agency, Vienna, Austria

[3] United Nations 2001 *Report on the Improvement of the Management of Radiation Protection Aspects in the Recycling of Metal Scrap* (United Nations Economic Commission for Europe, United Nations, New York, USA)

[4] Barton J, Hatton C and McMillan J 1991 *J Phys G: Nucl Part Phys* **17** 1885–1899

[5] McMillan J 1990 *Nuclear instrumentation for a superheavy-element search and for cosmic-ray studies* Ph.D. thesis University of Leeds

[6] Caines P 1972 *The development of an efficient neutron detector for cosmic ray studies* Master's thesis University of London

[7] Davidson P 1977 A new discriminator principle for slow neutron counting Tech. Rep. RL-77-106/A Rutherford Laboratory, Chilton, Oxon, UK

[8] Ziegler J, Biersack J and Ziegler M 2008 *SRIM – The Stopping and Range of Ions in Matter* (Lulu Press, Morrisville, NC, USA)

[9] Alburger D 1952 *Rev. Sci. Instrum.* **23** 769

[10] Gatti E *et al.* 1952 *Nuovo Cimento* **9** 1012–1021

[11] Zunger A, Katzir A and Halperin A 1976 *Phys. Rev. B* **13**(12) 5560–5573

[12] Waters L 2002 MCNPX user's manual: Version 2.3.0 Tech. Rep. LA-UR-02-2607 Los Alamos National Laboratory, NM, USA

[13] McMillan J 2015 The fabrication of wavelength shifting lightguides from clear acrylic sheet by disperse dyeing These proceedings

[14] Marsden E 2012 *Large area thermal neutron detectors for security applications* Ph.D. thesis University of Sheffield

The fabrication of wavelength shifting lightguides from clear acrylic sheet by disperse dyeing

J E McMillan[1]

[1] Department of Physics and Astronomy, The University of Sheffield, Sheffield, South Yorkshire, S3 7RH, UK

E-mail: j.e.mcmillan@sheffield.ac.uk

Abstract. Wavelength shifting lightguides have found extensive use as a means of collecting scintillation or cherenkov light from large areas onto a smaller area photodetector and for matching the emitted spectrum to the spectral response of the photodetector. Conventionally, such lightguides are fabricated by casting acrylic polymer with the fluorescent dye incorporated in the bulk. A technique has been developed in which plain cast acrylic sheet is disperse dyed in an aqueous bath. The resulting lightguide has the fluorescent dye held in a thin layer at the surface of the material. A number of different fluorescent dyes are demonstrated.

1. Introduction and inspirations

Wavelength shifting lightguides are extensively used as a means of collecting scintillation or cherenkov light from large areas onto a smaller area photodetector and for matching the emitted spectrum to the spectral response of the photodetector. Their operation is reviewed by Keil [1]. Conventionally, such lightguides are fabricated by casting polymethylmethacrylate (or polyvinyltoluene) with the fluorescent dye incorporated in the bulk. This inevitably requires tonne scale production. The technique suggested here is useful for small scale or bespoke production with experimental dyes.

Viehmann and Frost [2, 3] identified that the fluorescent dye does not need to be uniformly distributed through the bulk of the lightguide, but instead can be applied to the surface as a thin film of lacquer without loss of efficiency. However, it has been found impossible to reliably and repeatably produce optically perfect lacquer layers on large area acrylic sheets.

Mufson and Baptista [4] successfully dyed the surface layer of cast acrylic sheet with the wavelength shifter TPB by spraying with a solution in methylene chloride, followed by surface heating to melt the acrylic. This was done without degrading the optical quality of the surface.

This information inspired a search for other acrylic dyeing techniques. A domestic industry was discovered in which acrylic jewellery and similar items were being produced by surface dyeing using dyes intended for fabrics [5]. This dyeing was performed in hot water using the "disperse dyeing" technique.

Further investigation revealed that this technique was developed and published by ICI in the 1940's [6, 7, 8, 9]. The technique is also described in a recent technical brochure [10] from Evonik, the current German manufacturer of Plexiglas. It is perhaps surprising that it has not previously been used with fluorescent dyes for lightguide production.

2. Disperse dyeing

Disperse dyeing [11] is a technique extensively used in the textile industry for the colouration of fibres which are hydrophobic, such as polyester, nylon and cellulose acetate. The dyes used are insoluble in water at room temperature and are only sparingly soluble at higher temperatures, however, they have high solubility in the textile fibres. They are present in the dyebath as a fine aqueous suspension in the presence of a dispersing agent. The water dissolves individual molecules of the dye and the fibres then absorb the dye molecules from the solution.

When applied to acrylic sheet, a dyestuff is required which is largely insoluble in the aqueous phase but readily soluble in polymethylmethacrylate acrylic. In addition to water, the dyebath contains finely divided powdered dye and ingredients known in the dyeing industry as "auxiliaries". These are a surfactant which forms micelles, which carry the dye molecules from the powder grains to the surface of the acrylic. A "carrier" which causes the surface layer of the acrylic to swell and open sufficiently for the dye to penetrate. Finally, a "levelling agent" which produces a spatially even distribution of dye. The exact method of operation of this last ingredient appears unknown. The dyebath is operated at as high a temperature as possible, close to boiling point. After the dyeing process the acrylic must be washed, then baked to remove residual water and carrier.

3. Results to date

Experiments were performed on blue-green emitting wavelength shifters with the intention of fabricating low cost panels for collecting light from large area zinc sulfide sheets. Some initial experiments have been performed with TPB and bis-MSB in an attempt to produce a deep UV wavelength shifter.

The dyebath recipes were developed by examining previous work [5, 6, 7, 8, 10] and substituting materials with currently available equivalents, then performing experiments on small test samples.

The choice of carrier is important. If a carrier is chosen which is too aggressive, the surface of the acrylic melts and becomes sticky, thus permanently degrading the optical finish. If the carrier is not aggressive enough the dye will not penetrate the acrylic. Suggested carriers included benzyl alcohol (phenylmethanol) and acetone (propanone). It was found that isopropyl alcohol (2-propanol) was adequate and preferable to some others when volatility was considered.

A number of different surfactants were tried but best results were obtained with Triton X-100. Many of the earlier recipes suggested acidifying the bath with acetic acid. Experiments were tried at various values of pH, with little change in result, so acidifying was discontinued.

The last ingredient, the levelling agent, was chosen to be lecithin. The literature contained various suggested materials with wildly different costs and availability. Lecithin was one of these and its availability as a food ingredient at low cost made it attractive. Sufficient was added to the bath to make it resemble a thin broth. The result was a very even dyeing of the samples, rather than patchy or speckled colouration.

By polishing one edge of the acrylic, illuminating the face and viewing the edge with a microscope, it was determined that the depth of the dyed layer was around 40μm.

4. Experimental procedure

Experiments were performed with test strips of polymethacrylate acrylic. It was discovered that not all acrylic is of identical composition. Material which is extruded is affected much more by the carrier and takes up dye much more readily than cast material. It is thought that this reflects the fact that the average molecular mass of cast material is much higher than that of the extruded material. Since the long term intention was to dye large optically perfect cast sheets, the test strips had to be cut from this material. It is possible that samples from different manufacturers will perform differently, but this has not yet been investigated.

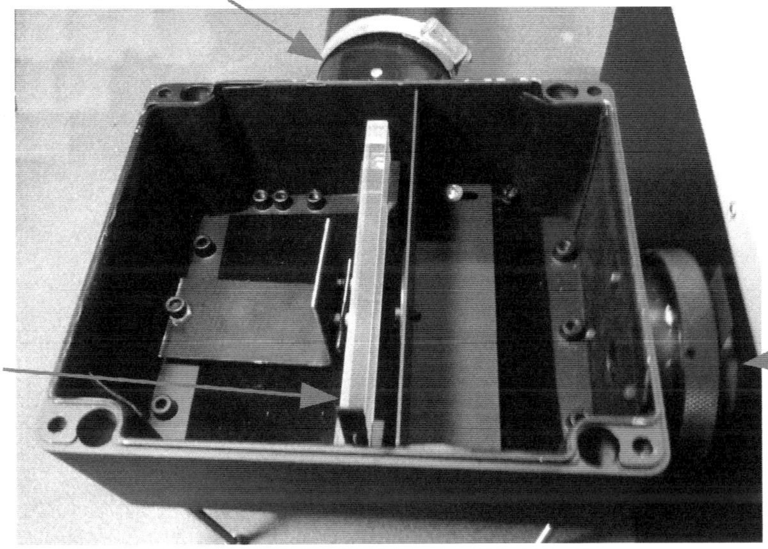

Figure 1. The experimental fluorometer

The test strips used were 190mm×40mm×10mm. These were treated in a volume of dyebath held in a re-sealable polyethylene bag suspended in a bain marie. A typical successful bath contained 300g water, 7g BBQ dye, 10g lecithin, 8g isopropanol and 2g triton X-100. It is not thought that any of these quantities were critical.

The samples were held in the dyebath for 30 minutes at a temperature of 95°C. By that point, further immersion did not result in greater uptake of dye. The samples were then taken from the bath, allowed to cool and washed in clean water for some minutes. They were then dried in an oven at 40°C for at least six hours.

Success has been achieved with BBQ (CAS 23749-58-8), Solvent Green 5 (CAS 2744-50-5), TPB (CAS 1450-63-1) and Bis-MSB (CAS 13280-61-0). Dyes which have high solubility in water, e.g. Fluorescein (CAS 518-47-8) and Solvent Green 7 (CAS 6358-69-6) were complete failures.

5. Initial optical measurements

In the initial stages, the dyeing process was optimized by altering the procedure and examining the test sample visually, deeper colouration being considered better. While this was a workable procedure for blue-green emitting dyes, it is completely unworkable with UV absorbing violet emitting dyes. Further, it is extremely subjective and objective measurements are clearly required.

Absolute optical measurements are notoriously difficult, however, much can be learnt from comparative measurements. A special purpose experimental fluorometer has been constructed, which is shown in figure 1. On the right hand side, a beam of light enters from an Optronic Laboratories OL750 spectrophotometer. The wavelength can be controlled from 200nm through to 1000nm. This light falls on the sample and produces fluorescence. A fraction of the fluorescent light passes down the sample and falls on a photomultiplier, which is operated in photon counting mode. Since the OL750 spectrophotometer does not produce equal intensities at all wavelengths, a calibration run against a Lambda UV/VIS radiometer is required.

As the wavelength of the incoming light is swept through the absorption band of the fluorescent dye, the photon count rate from the photomultiplier rises and falls. The output measures the integral of the emission spectrum of the dye convoluted with the spectral response of the photomultiplier. Better samples produce a higher rate, while worse samples produce a lower rate. Initial measurements have been performed which suggest that the surface dyed materials are comparable with commercially available waveshifter materials.

As the dye is not uniformly distributed throughout the matrix of the material, it is expected that the attenuation length will be long. An additional system is being planned which will measure the attenuation length of large samples.

6. Future work

When the optical measurements are completed, a more complete paper will be published. It is also intended to extend to other dyes, particularly those that fluoresce when exposed to deep UV light. There is also the possibility of using the technique to transport scintillating molecules such as p-terphenyl or naphthalene into the surface of acrylic to produce alpha particle sensitive detectors which are largely insensitive to beta or gamma radiation.

7. Acknowledgements

This work was done as part of STFC CLASP award ST/K000179/1. I would like to thank Lily Kamper and Jessica Sherriff for their advice on dyeing and the staff of the British Library for retrieving the ICI literature.

References

[1] Keil G 1970 *Nucl. Instrum. Meth.* **87** 111–123
[2] Viehmann W and Frost R 1979 *NIM* **167** 405–415
[3] Viehmann W 1980 Fluorescent radiation converter US Patent 4262206
[4] Mufson S and Baptista B 2013 *Journal of Instrumentation* **8** C09006
[5] Packington S 2010 *Making and Dyeing Acrylic Jewellery: A Full-colour, Step-by-step, Photographic Guide to Making and Dyeing Your Own Acrylic Jewellery* (How2crafts, Birmingham)
[6] Hill F and Simms E 1944 Colouring of articles British Patent 565942
[7] ICI 1945 The dyeing of preformed polymethyl methacrylate of the type of 'Perspex': The use of aqueous dyebaths, Technical Circular, Imperial Chemical Industries Ltd., London
[8] ICI 1948 The surface dyeing of Perspex acrylic sheet, Technical Circular, Imperial Chemical Industries Ltd., London
[9] Deverell-Smith R 1949 *J. Soc. of Dyers and Colourists* **65** 328–332
[10] Evonik Industries 2008 Surface treatment of Plexiglas: guidelines for workshop practice Ref. 311-4, Evonik Röhm GmbH., Darmstadt, Germany
[11] Broadbent A 2001 *Basic Principles of Textile Coloration* (Society of Dyers and Colourists, Bradford, West Yorkshire)

AUTHOR INDEX